An Anthropology
of the Machine

An Anthropology of the Machine

Tokyo's Commuter Train Network

MICHAEL FISCH

The University of Chicago Press   Chicago and London

The University of Chicago Press, Chicago 60637

The University of Chicago Press, Ltd., London

© 2018 by The University of Chicago

Published 2018

Printed in the United States of America

27 26 25 24 23 22 21 20 19 18   1 2 3 4 5

ISBN-13: 978-0-226-55841-7 (cloth)

ISBN-13: 978-0-226-55855-4 (paper)

ISBN-13: 978-0-226-55869-1 (e-book)

DOI: https://doi.org/10.7208/chicago/9780226558691.001.0001

Library of Congress Cataloging-in-Publication Data

Names: Fisch, Michael (Anthropologist), author.
Title: An anthropology of the machine : Tokyo's commuter train
    network / Michael Fisch.
Description: Chicago ; London : The University of Chicago Press,
    2018. | Includes bibliographical references and index.
Identifiers: LCCN 2017054085 | ISBN 9780226558417 (cloth :
    alk. paper) | ISBN 9780226558554 (pbk. : alk. paper) |
    ISBN 9780226558691 (e-book)
Subjects: LCSH: Railroads—Japan—Tokyo—Commuting traffic. |
    Urban transportation—Social aspects—Japan—Tokyo.
Classification: LCC HE5059.T6 F57 2018 | DDC 388.4/20952—dc23
LC record available at https://lccn.loc.gov/2017054085

*This book is dedicated to my partner Jun
and to our sons Kai and Mio.*

Contents

Preface

This book is a technography of collective life constituted at the interplay of the human and the nonhuman, of nature and machine. Its central scene is Tokyo's commuter train network, one of the most complex large-scale technical infrastructures on Earth, where trains regularly operate beyond capacity. This book treats this scene both as an articulation of specific sociohistorical relations between humans and machines and as a general expression of a current but also potential condition of collective life. The weight of analysis falls on the latter, the potential of collective life, for this book is an argument concerning not only what collective life *has* become but moreover *what it can become* under contemporary conditions of media and technology.

The events of March 2011, when a strong earthquake off the northeast coast of Japan sent a massive tsunami into the shore, killing thousands of people and causing several reactors at the Fukushima nuclear-power plant to melt down, have imparted urgency to the question of how we might inhabit and collectively survive within current and future socio-technical conditions. The tragedy of March 2011, or 3.11 as it is known in Japan, resists neat categorization as a human, technological, or natural disaster. It was all three simultaneously, demonstrating the absolute meaninglessness of a mode of thinking that remains confined to bounded sets of relations. This book takes up the challenge of rethinking technology by examining a large-scale transport infrastructure in Japan, where the issues provoked by 3.11 are inhabited in a daily and regular manner, and where we can begin to develop an anthropological media theory of scale and ecology.

In taking this approach, this book advances the normative claim that we need to transform our understanding of technology if we hope for collective life to not only survive but thrive on this planet. Just as there can be no collective future without technology, there will certainly be no future collective without a significant transformation in how we think of technology and what we demand of ourselves in relationship with it. This is not a claim that technological development will save life (human and nonhuman) on this planet. Rather, this book is an argument for a different kind of ontological entanglement with technology, one that stresses a dynamic quality of ethical relationality and trust instead of rationalized interactions and profit. By *relationality*, I mean a system that is less rather than more determined, a system that has increasing leeway for interacting, thinking, and becoming with the human and nonhuman environment. The term I use to capture the notion of a relationality of quality and trust with technology is *technicity*. Coming out of a long history of machine theory, *technicity* denotes a machine's degree of dynamism and openness to current and future relational flourishing or becoming. The term emphasizes a technology's ontological and conceptual affordances, as well as its trustworthiness as a partner of collective life in the present and future. This is not an argument for human exceptionalism. It is, rather, an argument for a post-human humanism that recognizes the equal importance of technology, human, and nonhuman in the formation of a robust collective life while placing exceptional responsibility on human beings to maintain the dynamic and diverse integrity of collective emergence. Technography is the medium of post-human humanism. Embracing an experimental, speculative modality, technography seeks to open collective futures.

Among the many things that Fukushima revealed is the woeful inadequacy of the term *technology* for parsing the complexity of our contemporary collective life. The term *technology* does not allow us to make ethical distinctions between such vastly different kinds of machines as nuclear reactors, commuter trains, and more mundane personal devices like smartphones. It flattens all these machines into a single category. Even aside from their obvious but significant differences in scale, it seems like common sense that these are incommensurable kinds of machines that engender vastly different kinds of relationships—yet we have no real way to talk about the quality of ontological entanglement these technologies allow. *Technology* merely denotes a value-free instrument, a means to an end, whose successful (read: "uneventful") operation is reduced to a matter of rational governance and technological management. This book rejects this reduction. It argues instead that we must begin to think about technology differentially, in terms of its trustworthiness. *Is tech trustworthy? Makes me think of all the fear of it taking over, stealing jobs.*

Technological trustworthiness is not only about reliability, resilience, and fail-safe mechanisms. Although these are important attributes for any machine, they are not necessarily what makes for an ethically oriented ontological entanglement. Hence the argument, which one often hears from advocates of nuclear energy, that good nuclear power is just a matter of better reactor design and more-rational systems of management does not enter into how this book formulates trustworthy technology. Machines with which we can be in a relationship are machines that can be in a relationship with us. Trustworthy machines do not *another very human quality.* demand compliance; they are forgiving and ontologically capacious in *interesting choice of words* their capacity to evolve with collective life.

Why develop such an argument through a train system, let alone Tokyo's commuter train network? Surely there are more timely techno-assemblages with which to think about current collective life, such as biotechnology, the internet, or even smartphones. Isn't a commuter train network merely an obvious instantiation of modern industrial technology and a bygone modality of value production through the capture of surplus human labor and attention? I contend that Tokyo's commuter train network is the ideal medium with which to rethink technology because the network operates beyond capacity and because generally we understand the train to be the originary machine ensemble in the evolution of modern industrial society and the advent of our current second-nature technological condition. In addressing questions concerning technology by thinking through Tokyo's commuter train network, this book develops a conceptual history of the train via technicity and opens up alternative ways for thinking about future collective life. This book posits that if we can tell the story of Tokyo's commuter train through its margin of indeterminacy and technicity while emphasizing questions of dialogue and relationality over tropes of conditioning and determination, then we can resist received technological narratives and identify novel limits and novel possibilities for trustworthy technologies of collective life.

Introduction: Toward a Theory of the Machine

Tokyo's commuter train network is a complex web of interconnecting commuter and subway lines that dominates the urban topography, providing the primary means of transportation for upward of forty million commuters a day from the city's twenty-three inner wards and three adjoining prefectures.[1] On a typical weekday morning, the system's ten-car commuter trains are packed at 175 percent to 230 percent beyond capacity. This means that a train car designed for no more than 162 people will carry between 300 to 400 commuters. With seven to ten people, rather than three, occupying each square meter of floor space, commuters are squeezed together so tightly that they can barely breathe.[2]

This operation beyond capacity defines Tokyo's commuter train network. Arms caught among the compressed bodies have been broken, and commuters sometimes lose consciousness for lack of oxygen. When they do, they remain standing, propped up by the collective pressure of the surrounding bodies. But operation beyond capacity is not just about train-car congestion. It concerns traffic density as well. During the morning rush hours, train operators must stream one train after another with the absolute minimum gap between them in order to accommodate the commuter demand. On main train lines, that gap is less than two minutes. Because of the relatively short distance between stations, the high-capacity and high-density traffic places enormous strain on the infrastructure, creating highly precarious conditions whereby a delay of any kind catalyzes a vicious cycle, leading to platform crowding and more delay that can spread quickly to train lines throughout the network, causing a systemic collapse of order.

Nowhere is the precariousness of operation beyond capacity more clearly expressed than in Tokyo's Yamanote Line, which circles the center part of the city. The Yamanote Line has twenty-nine stations, only 1.5 kilometers apart on average, and is linked to every major train line.[3] During the morning rush, the typical ten-car train on the Yamanote Line carries between three and four thousand commuters (three hundred to four hundred people per train car), and traffic operates with just under 2.5 minutes headway between trains. If a train is delayed five seconds at each of the twenty-nine stations, the cumulative effect over the course of the twenty-nine stations is 2.5 minutes, which is equivalent to the minimal interval between trains. Consequently, a ten-car train must be cancelled to make space, leaving three to four thousand people with no choice but to try to cram themselves into the remaining trains. The effect is an inevitable delay that proliferates through the system. Operation beyond capacity, it would thus seem, demands an absolute and tightly coordinated schedule that does not allow for any divergence.

Paradoxically, during the Tokyo commuter train network's morning rush, not only are trains regularly delayed without the system collapsing into disarray, but also, more importantly, those delays allow for operation beyond capacity. Even at the line's most crowded stations (Shinjuku and Shibuya), the allotted dwell time (stopping time) for boarding and debarking is a mere fifty seconds. At other stations, the allotted dwell time is only thirty seconds. As platforms fill with long lines of commuters, it is simply impossible for commuters to exit onto the crowded platform and for waiting commuters to squeeze into the filled trains in that short time. Train drivers are thus forced to extend dwell time at stations, sometimes by as much as half a minute, in order to accommodate commuters. Lost time must subsequently be recovered or partially recovered. Ten seconds lost to extending dwell time at a station can be recovered in the interval before the next station by applying slightly more acceleration on departure and waiting until the last possible moment to brake when entering the next station. Recovery from an even greater temporal deficit, however, presents a more considerable and often impossible challenge. A thirty-second delay might be recovered entirely in the interval between a number of stations. Most likely, though, it will not, which forces operators to recalibrate and

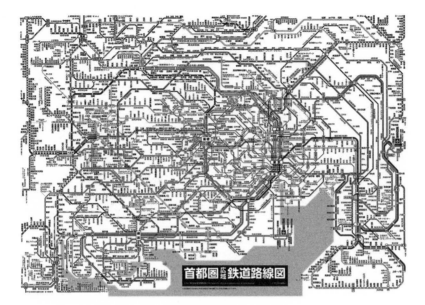

FIGURE 0.1. Tokyo Commuter Train Network Map
Source: CHIRI Geographic Information Service

tweak the gap on other train lines. Operation beyond capacity is thus realized through carefully managed divergence from the scripted order, not adherence to absolute punctuality.

For train drivers and system operators, managing divergence demands close attention to the gap between the specified order and the actual performance. Every second counts and every second is accounted for. For commuters, managing the gap is the technique of commuting, eliciting a constant, embodied, and active attention to the network's fluid order. Even during off-peak hours, it remains the guiding principle of train traffic management.

Crowded commuter trains are a facet of everyday life in urban centers throughout the world. In this book, I posit that the specificity of Tokyo's commuter train network lies in the significance of the gap for operation beyond capacity. The gap is schematized in the train-traffic diagram, or *ressha daiya* (hereafter just *daiya*). A traffic diagram is a universal technology for planning and managing a schedule within a restricted transportation system. Railroads and airports use traffic diagrams; expressways and highways do not. In a railroad-traffic diagram, the movement of trains is plotted on a horizontal axis of time and a vertical axis of stations.[4] Each train line has its own traffic diagram,

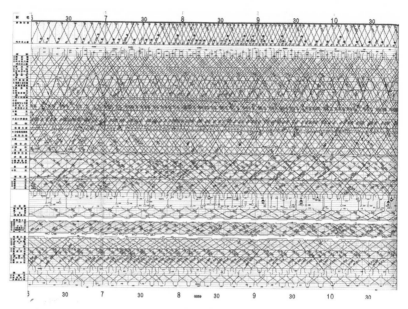

FIGURE 0.2. Principal *daiya* for train traffic between 6 a.m. and 11 a.m.

and each line of the diagram represents a single train, with the angle of the line indicating the specified speed of the train on sections of track: the more vertical the line, the faster the speed, and the more horizontal the line, the slower the speed. In addition to providing schematics of the actual track layout, environmental conditions such as slopes and curves, and signal locations, a train-traffic diagram specifies the different types of service (local, express, semi-express), different train technologies, and allotted stopping times at stations. In sum, the traffic diagram determines the spatiotemporal order of the commuter train network in addition to providing everything railroad operators need to know in order to operate the train.

The *daiya* is produced by expert technicians known as *sujiya*. For most of Japan's postwar period until the mid-1980s, the nation's main rail-transport company, Japanese National Railway, recalculated and re-drew *daiya* annually during monthlong retreats for *sujiya* at secluded hot-spring resorts.[5] Today, much of that process has been eliminated, and the *daiya* can be redrawn several times a year with the assistance of computers and the compilation of commuter data constantly mined from electronic ticket gates in train stations. When *sujiya* and infor-mation scientists talk of recalibrating *daiya*, they speak in terms of optimizing traffic patterns in correspondence with shifting trends in ridership discerned from this data. Their objective is to create a more "convenient *daiya*" (*benri na daiya*) that more accurately reflects the lived interaction of commuters and the commuter train network and also anticipates emergent commuter needs. Although *daiya* do not circulate among commuters, they command their constant attention. Everyone knows of commuter-train *daiya*, and everyone remains perpetually aware of the state of the *daiya* for their particular commuter line via posters in the trains that keep them informed of upcoming *daiya* revisions (*daiya kaisei*). Morning television and radio news programs also provide regular *daiya* updates.

While commuters in Tokyo tend to think of the *daiya* for each train line as a single, determined object not unlike a schedule, it is actually a combination of two components. There is a planned "principal" (*kihon*) *daiya*—a painstakingly calculated, idealized configuration of traffic flow—and an actual "operational" (*jisshi*) *daiya*, which emerges in accordance with the overall fluctuating circumstances of actual train operation.[6] Whereas the planned *daiya* refers to the temporality of clock time and delineates a schedule, the actual *daiya* reflects the lived tempo of the city and train network. This two-part composition lends the *daiya* a dynamic quality that leads technicians and system operators to call it "a living thing."[7] → more humanization.

Operation beyond capacity in Tokyo's commuter train network depends on maintaining the gap between the two *daiya*, making the gap a central focus of system operators. At the same time, operators must tend to more than just the gap between the *daiya* of a single train line. The dense and interconnected nature of the system demands a more global attention to the overall network condition, which is an expression of the correlate gaps of all of its train lines. In this book I borrow from the French philosopher and machine theorist Gilbert Simondon in identifying the field of interaction constituted by those correlate gaps as the Tokyo commuter train network's "margin of indeterminacy."[8] The network's margin of indeterminacy, this book posits, is its dimension of collective life. It is a domain of ontological entanglement where the processes of humans and machines intersect with the time and space of institutionalized regularities to produce a provisionally stable techno-social environment of the everyday. While operation beyond capacity in Tokyo's commuter train network is a mode of technological organization inseparable from the social and historical conditions of Japan in general and Tokyo in particular, it also makes legible the margin of indeterminacy as a principal quality of technical ensembles.

Important paragraph?

This book situates itself within the margin of indeterminacy of Tokyo's commuter train network. By thinking with the processes, practices, tensions, and contradictions articulated within the margin of indeterminacy, it develops a technography of the commuter train network. In so doing, it forges a *machine theory* adequate to the experiences, practices, and ethical questions that emerge within the immersive technological mediations that define contemporary collective life. Technography takes its cue from ethnography as the time-honored method of anthropology for generating analytical interventions into human society through detailed descriptions of specific human practices and modes of social organization.[9] But in replacing *ethno-* with *techno-*, technography works to accommodate a growing consideration within anthropology for cultures and practices of technological mediation that are irreducible to categories of identity, community, nation, agency, and subjectivity. Technography, I insist, must also move beyond anthropology's representational mode of knowledge production. Merely describing a technological condition or in situ processes whereby people adapt technologies to realize a specific outcome does not suffice. Not only does such a descriptive approach risk reifying a binary structural ontology of human versus machine, it is also burdened with a problematic, twentieth-century anthropological conceit for producing knowledge of an other. A technography must instead become performative by thinking *with*, not just *about*, technology. Such an approach is in concert with the call for empirically driven theoretical thought born of an encounter with material and immaterial conditions.[10]

The machine theory for thinking with technology that I develop in this book derives from the margin of indeterminacy of Tokyo's commuter train network. Its goal is to think with the processes of immersive technological mediation and conditions of human-machine interaction. Commuters emerge within these processes not as subjective positions constituted in opposition to the system's technology and the corporate enterprise behind it, but rather as iterations of a collective distributed across a technologically mediated milieu.

Not too long ago, a technography of a commuter train network would have been a profoundly difficult proposition.[11] Large technological infrastructures were simply not conventional sites of anthropological inquiry.[12] What is more, anthropology had yet to develop a robust theoretical orientation toward technology that could move beyond a concern with technology's relationship to modernity and its perceived mechanizing effects on the social mind and body.[13] Much has changed in recent decades, primarily as a result of scholarly work in the field of

Assemblages?

science and technology studies (STS) making tremendous inroads into the study of infrastructure, media, and technology.[14] While I draw on this literature throughout this book, I take my main inspiration from neither STS nor anthropology; rather, I turn to a group of "machine thinkers" who emerged after World War II, at the height of cybernetics, whose work articulates the initial tenets of what I call *machine theory*.

Simondon is at the center of this group of thinkers. Writing mostly in the 1950s and 1960s, Simondon was concerned with the evolution of the collective formed in the interplay between humans and machines. He developed a unique approach to technology, one aimed at overcoming what he saw as an opposition between culture and technics that had led to a reductionist, utilitarian approach to machines.[15] In Simondon's thinking, machines are more than tools external to an ontologically stable human subject; rather, they are integral to the processes of human thinking and social becoming. Simondon's work had a deep impact on thinkers of the time, especially Gilles Deleuze. Nevertheless, only in recent years have his writings begun to receive close attention from scholars in philosophy, technology, and media. These scholars have extrapolated his ideas to think with today's increasingly sophisticated and ethically complex technologies and technical ensembles. I draw on this expanding body of work for my own thinking with Tokyo's commuter train network.

Like many machine thinkers in the early postwar period, Simondon developed his theories through an engagement with cybernetics. The impact of cybernetics in producing our present conceptual and material reality cannot be overstated. Although we typically associate cybernetics with the emergence of information theory, Cold War infrastructure, and research into artificial intelligence (which dissipated with the loss of funding in the early 1970s), a number of scholarly works in recent years have emphasized the profound impact of cybernetics, as an international and interdisciplinary project, in changing the direction of thought and practice in everything from architecture to philosophy, social theory to economic theory, financial systems to governmental rationality.[16] The media historian Orit Halpern does not exaggerate when she argues that cybernetics restructured how we encounter the world by reshaping our perception, rationale, and logic.[17]

Postwar machine theory embarks from the impulse in cybernetics to move beyond the dualistic presuppositions of technological determinism that have fueled either naive visions of techno-utopias or anxiety over the loss of human autonomy to a machine master. Such thinking dominated the discourse concerning the impact of machines on

[handwritten marginal note:] Machines don't just help us; they quite literally change the way we see things in the world and live in the world and of us.

human society throughout much of the twentieth century and can still be found in many popular mainstream forums today. Machine theory, by contrast, perceives the relationship between humans and machines in dialogic terms. Its idioms are coconstitution and interaction rather than dominance and control. At the same time, it rejects the fundamental proposition of cybernetics espoused by cybernetics' originator, the American scholar Norbert Wiener, that all life is reducible to information processing.[18] For scholars like Simondon, the reduction of life to information processing gives precedence to form over matter and is indicative of the dualistic and functionalist thinking that informs technological determinism. Machine theory, by contrast, treats information on an ontological level as an "intensity" and a material force.[19]

Similarly, machine theory rejects "cyborg" metaphors of human and machine symbiosis.[20] Its notion of technical becoming is not about the fusion of human and machine. It insists instead on a fundamental, ontological incommensurability between human and machine, and it insists on maintaining a space of difference as that which animates both. Maintaining this space of difference is critical for thinking about the relation between commuter bodies and the commuter train network in this book. Tokyo's commuter train network is not a platform for cyborg humanism; it is a scene of collective life constituted in the interplay of humans and machines that poses questions concerning the limits and potentials of our current technological condition.

Although I draw a distinction between machine theory and STS, in many respects STS cannot be separated from machine thinking. Indeed, many fundamental concepts in STS draw on the nondualist, nonsubstantialist approach to the interaction of things as co-constitutive processes (not identities) that was developed by postwar machine thinkers. This approach is explicit, for example, in Bruno Latour's attempt to move beyond the discursively constituted ontological boundaries separating human and machine in order to represent the active, agentive role of nonhuman things in the formation of a collective life.[21] While it has helped Latour transform what we think of as the social into a far more capacious, contingent, and processual collective of human and nonhuman actors (which includes everything from bacteria to objects, machines, and infrastructure), it has also guided feminist STS-related thinkers such as Donna Haraway in calling attention to the relational ethics of collective processes.[22] Emphasizing a commodious notion of collective, both STS and machine theory are invested in ontological questions, exploring how technology and technical things perform as material forces irreducible to symbols and representation.

Machine theory thus complements efforts among STS thinkers to shift ethnographic methods from a representational to a performative mode of engagement whereby the ethnographer is enfolded into the generative processes of contextualized practices and materialities.[23]

While much of STS is concerned with questions of knowledge production and technological practice, machine theory retains the speculative utopianism of cybernetics as it advocates the possibility for a different kind of relationship with technology. Machine theory asks not just what technology is and how it impacts or draws together social relations but also—and more importantly—how it works, what it does, and what it might become. Machine theory is invested in thinking *with* technological ensembles toward novel conceptual formulations while simultaneously maintaining a critical perspective on the kind of collective technology has enabled thus far.

With its focus on the ontogenetic affordances of technology, machine theory also departs from conventional forms of capitalist critique. Machine theory asserts that humanity's problem rests primarily in its relationship with machines, not in the logic of capital and its corollary discursive structures. Simondon parsed this as a matter of alienation. In contrast to Marx's theory of alienation as an effect of the structure of labor under capitalism's relations of production, Simondon understood the problem to be humanity's alienation from the machine and argued for an unconventional humanism in which human society would realize a collective potential through "technical becoming."[24] Simondon thus rejected the privileged status that social theory afforded to labor as the singular authentic site of social becoming, rather placing his hope on the notion of a novel collective becoming, one born of heightened "technical attitude."[25]

Guided by this approach, this book departs from historical analyses that have explored the development of the commuter train network in Japan as an exemplary instantiation of urban development under capitalism.[26] Situating itself within the margin of indeterminacy of Tokyo's commuter train network, this book asks how the tensions and contradictions that form under operation beyond capacity urge us to think, imagine, and practice toward novel forms of ethically bound collective becoming. In so doing, this book attends to the genesis of the schema of operation beyond capacity and its contemporary instantiation in Tokyo's commuter train network as a mode of techno-social organization. It also looks to re-mediations of the network's tensions and contradictions in films, advertisements, and web-based social media. Overall this book focuses on the processes of human and machine interaction within conditions of immersive technological mediation that constitute the collective life of Tokyo's commuter train network.

The Modern Machine

Developing a technography of a train network requires dealing with theoretical baggage concerning the train as the historical mainspring of modern industrialism and the central driving force in the rationalization of human society. The historian Wolfgang Schivelbusch encapsulates this notion with the subtitle to his history of the railroad's development in Europe and the United States: "The Industrialization of Time and Space in the Nineteenth Century."[27] Schivelbusch's text has become a seminal work for scholars interested in technology and society, and it is typically read as a generalizable narrative of technological development in capitalist modernity. Its argument builds on an understanding, laid out in the initial chapters, of the train as the first "machine ensemble," by which Schivelbusch means the first expansive assemblage of technological components, systems, and subsystems whose seamless interaction was necessary for operation without disaster. In other words, the railroad was not just a machine: it was the first iteration of an emergent machinic ecology. Extending far beyond the basic tracks and stations, this machinic ensemble transformed the topography of the land while giving rise to a network of tightly coordinated auxiliary and affiliated industries ranging from the coal mine to the factory, the publishing house to the department store, the bed town to the resort town. Schivelbusch thus shows how the railroad was a driving force in the emergence of a technologically engineered environment that became the forerunner to the condition of immersive technological mediation of contemporary society.

In Schivelbusch's argument, the railroad effects the displacement of rhythms, views, and experiences of a premodern natural world with the tempos, pathologies, and sensations of a constructed, technological environment. Accordingly, this subjection of the natural world to the machine transpires in conjunction with the rationalization of the human sensorium and of human social relations. Veering at times toward technological determinism, Schivelbusch depicts the railroad as retooling human perception, thought, and social behavior in correspondence with the mechanical speed of the train, its schedules, and its operational imperative to engender a modern industrial experience of time and space. Nowhere is this clearer than in Schivelbusch's description (drawn from Georg Simmel) of passengers learning to manage the awkward intimacy imposed on them by the tight quarters of the train car and the development of a novel panoramic vision of the passing landscape perceived through mechanized speed.[28] Ultimately, Schivelbusch's account becomes a story of technological development as the loss of a premodern, nontechnological sensibility. This loss occurs in conjunction with intensified social rationalization, as the speed and complexity of the technological apparatus necessitates a heightened degree of technological efficiency and disciplined passenger behavior.

Whether situated in Europe, America, or Japan, stories about trains and histories of the advent of the railroad tend to follow in Schivelbusch's tracks, depicting the arrival of the train as fueling the rise of industrial society and the corollary struggle for authentic human relations against an oppressive rationality and automaticity of machinic life.[29] The train performs thus as the vanguard of capitalism's rationalizing, machinic logic, crushing the organic character of premodern social relations beneath its unforgiving steel wheels as it effects the mechanistic conditioning of minds and bodies toward the formation of a mass-mediated modernity. Similarly, the commuter train figures as a powerful vehicle of capitalist alienation that subjects time, space, and bodies to the merciless logic of capital as it transforms landscape into real estate and mediates transitions from home to work and school.

Through such theoretical expositions of the train, which anchor narratives of the historical shift from premodern to modern, trains and commuting have become bound up in the ideological mediations of modern technological infrastructure—mass transportation, mass production, and mass media. At the same time, these theoretical expositions insist on an intractable logic, perhaps best explicated in Georg Simmel's description of the clockwork relations of the early metropolis, whereby technological development that was initiated under the steam engines of the late nineteenth century leads to increasingly complex and tightly coordinated interaction between humans and machines.[30]

This story of the train becomes paradigmatic of the human relationship with machines in modernity. It has fueled dystopian prophecies warning of the automaton-ization of human society as well as cathartic visions of engaging in total war with machines in order to save the human race from machinic enslavement or extermination. A somewhat more optimistic but similarly invested approach underscores moments of machinic excess—points of shock, disruption, and instability—as potential sites of redemptive aesthetics and irrationality. Such points of excess are then celebrated as the condition of possibility for the recovery of something human, generally in the form of stories of romance, crime, and intrigue against a technological background.

It is not difficult to align Tokyo's commuter train network with this narrative. Indeed, the network's famous precision and its infamous spectacle of fantastically packed commuter trains suggest a population of mechanistically conditioned commuters yielding to rationalizing technological forces—"trained," as it were, to the operational imperatives of the apparatus. Accordingly, the packed commuter train easily figures as a spectacular expression of capitalism's rationalizing logic whereby human beings are objectified as mere cargo, conveyed in accordance with the merciless dictates of mass production.

As compelling as this narrative may be, it represents a significant reduction of the historical and local complexity at work in the experiences of technological mediation that are part of the commuter train network. It also leaves us with nowhere to go theoretically but off the train, which becomes an especially problematic move for the way it constitutes a romanticized ideal of either a pre-technologically-mediated past somewhere outside the commuter train network or a digital, post-industrial, and postmodern future. Alternatively, we can try to stay on the train while resisting its subjugating force by insisting on recovering some persisting human essence that escapes technology's colonization, which is precisely the strategy of so many twentieth-century films and novels that employ the train as a mise-en-scène of human drama.

The initial challenge of this book can be summarized as the question, *How can we stay on the train and engage directly with its scene of technological mediation, in ways that attend to the historical and situated specificities of its human and machine relations, so as to open new possibilities for thinking about modes of collectivity and technological becoming?* This is the question of machine theory. In other words, how can we come back to the train through an understanding of its technological condition in ways that escape the teleological discourse of machinic modernity and its inevitable effects? How can we think with the train rather than simply invoking it again and again as an exemplification of technology's mechanistically rationalizing processes under capitalism?

Theory from the Gap

When the media studies and Japan scholar Thomas LaMarre revisits Schivelbusch's thesis in the introduction to his work on Japanese anime, he does so through a machine theory that draws on such thinkers as Gilbert Simondon, Gilles Deleuze, Félix Guattari, and Martin Heidegger.[31] In so doing, LaMarre encourages us to reread Schivelbusch not as an allegory of modernity and mechanistic conditioning, but rather (in the vein of cybernetics) as an explication of a novel, immersive, mediated feedback environment that lends itself to different modes of thinking and becoming with the machine. Via Schivelbusch, the train in LaMarre's work becomes a technology that is good to think with. In contrast to Claude Lévi-Strauss's famous emphasis on thinking with animals toward the exposition of a structural model of symbolic associations, for LaMarre, thinking with the train emphasizes an ontological engagement in line with Simondon's approach to technology.[32] The train asks us to think with its material intensities in order to develop analogies for thinking about the conditions of immersive technological mediation.

Whereas the narrative of technological modernity posits the railroad as effecting increasingly hermetic and rationalized relations, LaMarre's thesis draws attention to points where Schivelbusch's argument asks us to understand the train as producing "gaps," or situations in which technologically-influenced perspective transpires as a material force, that elicit new forms of experience and thought. What is more, LaMarre emphasizes the emergence of a gap at the most totalizing moment of immersive technological mediation: when the train passenger's vision of the world outside the train becomes mediated by the novel experience of mechanized speed. If, for Schivelbusch, the passenger's experience of speed-blurred vision from the train window instantiates a split from a premodern panoramic vision, then, for LaMarre, perception at mechanized speed is important for the gap it generates. As LaMarre writes, "speed introduces a new kind of gap or interval into human perception of the world, and that specific interval, that manner of 'spacing,' does not serve to totalize the whole of perception or of experience related to train travel. Rather, the new interval or spacing folds humans into its operation and starts to rely on other machines such as printing presses, department stores, and carriages or cars."[33] In accordance with machine theory, the gap is a phenomenon of ontological significance. It is a space in which perspective transpires as a material force that elicits novel organizations of becoming with the machine. Such becomings incorporate the possibility of a new relationship with technology that unfolds from within the protean gaps, intervals, or spacings. In LaMarre's work gaps emerge as "zones of autonomy" that elicit new practices, new activities, and new ways of perception and cognition. *Are we the machine?*

In identifying gaps, intervals, or spacings as sites of ontologically driven, conceptual individuation, LaMarre encourages us to approach the evolution of a technological ensemble such as the railroad from the perspective of its generative relations rather than its deterministic effects. This is not just about identifying a causal relation between the emergence of new machine ensembles and novel, technologically driven practices—for example, commuting gives rise to reading on the train, which gives rise to the corresponding publishing industry, and so on. Rather, it is about recognizing potential phenomena of co-constitution in the evolution of technical ensembles whereby the emergence of zones of indeterminacy elicit an interweaving of machinic and corporeal processes toward novel conceptual interventions. Under this approach, a technical ensemble potentially becomes a kind of "thinking machine," which LaMarre defines as a "heteropoietic process [involving human and machine] in which human thinking happens differently than it would otherwise, in another flow of material forms and immaterial fields."[34] Following LaMarre, throughout this book I will at times use such phrases as "thinking the train" or "thinking the train with the web" to gesture to this kind of heteropoietic process.

In Simondon's work, a technical ensemble's "margin of indeterminacy" is the scene of collective co-constitution and is ultimately, I argue, where it becomes a "thinking machine." In contrast to LaMarre's formulation of the gap, the margin of indeterminacy in Simondon's machine theory performs a more functional role. Nevertheless, I argue that it provides an underlying condition of possibility for the emergence of various forms of gaps, intervals, and spacings that I will identify throughout this book in relation to the co-constitutive processes of Tokyo's commuter train network. In the simplest sense, the margin of indeterminacy of a technical ensemble denotes the openness the ensemble maintains internally to external information and thus is what allows it to incorporate the changes and contingencies of its environment into its pattern of operation. As such, the margin performs as a structurally underdetermined zone of interaction between the ensemble and its environment, allowing for the resolution of conflicts between the internal organization of the technological ensemble and external forces. In a work that mobilizes Simondon toward a reassessment of technology, the media theorist Adrian Mackenzie aptly paraphrases the significance of this zone of indeterminacy when he writes, "A fully determined mechanism would no longer be technological; it would be an inert object, or junk."[35] As Mackenzie suggests, the margin of indeterminacy is what allows the technological ensemble to suspend

[handwritten margin note: How machines shape our thinking]

"final determination of its own form" and to remain continuously in formation and able to incorporate variation.[36] Junked technology, like a bricked iPhone, is a machine that is unresponsive to input.

But why does Simondon use the somewhat cumbersome phrase *margin of indeterminacy*? Why not just call it a technology's operational latitude, or its viable pattern of divergence from normal operation? The term *indeterminacy* traverses two fields of practice central to Simondon's thinking: quantum physics and philosophy. In both, it carries important ontological connotations that set it apart from terms like *latitude* or *uncertainty*, the latter of which has more to do with risk and unpredictability. *Indeterminacy* refers specifically to the incompleteness of an individual, by which Simondon means not the juridical or philosophical notion of personhood and subject, but a provisionally stable set of functional relations deriving from an environment. Individuals, in Simondon's thinking, can be physical things (rocks or simple tools), biological organisms (humans, animals, bugs, or trees), and/or machines (engines or commuter trains). Incompleteness is not about a lack. It emphasizes, rather, the irreducibility of an individual to an inherent identity, essence, or substance.[37] By remaining incomplete, an individual remains open to "information" and subsequently to further transformative interactions with a milieu.[38] Incompleteness is thus the condition of possibility for ontogenesis, or material foldings from which new functional associations that develop into sedimented patterns of interaction emerge. The margin of indeterminacy thus becomes the scene of the processual ontological entanglement of humans and machines—the scene of collective life. Again, assemblages?

The Margin of Indeterminacy and Large-Scale Infrastructure

As the scene of technical ensembles' collective life, the margin of indeterminacy provides an avenue for anthropological inquiry into a large-scale infrastructure such as Tokyo's commuter train network. What is more, it does so in a way that overcomes a number of theoretical and methodological challenges posed by the complex ontology of large-scale technical infrastructures.[39] Embodying contingent political histories, technological shifts, and expert knowledge, large-scale technical infrastructures make for fascinating but also unwieldy and unyielding sites of anthropological inquiry. They can be mediums of time and space that traverse borders, cities, and culturally distinct regions to allow for the movement of people and things; at the same time, they can be a form of a place (or "non-place") with its own particular spatiotemporal character.[40] Designed to discourage dwelling and enable mobility, large-scale, technical transport infrastructures tend to be inhabited by indeterminate publics that are temporally manifest in various forms of communication and technologies.[41] Finally, large-scale technical infrastructures tend to be systems of systems formed in a fusion of mechanical, electrical, and informational technologies that ground our daily lives but often remain unseen and unacknowledged. The result is a messy techno-social topology that resists attempts to bring the pieces together in a nonreductive and coherent frame of analysis. The problem is not about finding an Archimedean point from which to constitute infrastructure as a proper, bounded object of inquiry: that would simply produce an ossified, lifeless structure. The problem, rather, is about developing an analytical orientation through which one can make legible the historically contingent material and immaterial processes of a large-scale infrastructure while simultaneously engaging the ethical problems raised within its particular articulation of collective life.[42] The margin of indeterminacy of a technical ensemble offers such an orientation.

The margin of indeterminacy of a large-scale technical infrastructure is not a place. It is a field of interaction and a medium for provisional resolutions of conflicts between the disparate processual orders of humans and machines. Such interactions make it a zone of perpetual tension marked by emergent processes that sediment via reiteration into temporary structures. While the margin of indeterminacy of a tool or machine can be fairly straightforward and easy to identify, Simondon contends that the margin of indeterminacy of a technical ensemble tends to be an extraordinarily complex aggregate of the correlate margins of indeterminacy of each of its parts.[43] For Simondon, this complexity makes the margin a particular kind of problem space. He writes that the margin of indeterminacy of a technical ensemble "cannot be calculated, nor be the result of calculation; it must be thought, posed as a problem by a living being and for a living being."[44] Simondon thus puts the human being at the center of the margin of indeterminacy, thinking with the constitutive tensions of its relations. But he also makes human beings accountable for the nature of the collective that is enabled in those relations. That is, posing the margin of indeterminacy as a problem becomes a matter of not only thinking with the margin of indeterminacy and its complex processes of human and machine interaction, but moreover calling attention to its "technicity"—its *quality* of relations. *Technicity* is a term that I will unpack and develop throughout this book in regard to the constitutive and emergent relations of Tokyo's commuter train network. It is an important term that carries ethical connotations, in that it asks us to think in terms of the degree to which a margin of indeterminacy allows further ontogenesis beneficial to both humans and machines and thus affords the flourishing of collective. *Technicity* is a term through which I think about the ethical integrity and quality of the collective specific to Tokyo's commuter train network.

Gaps, Spaces, Intervals, and Margins

To live in Tokyo is to live on and by the commuter train network. Every morning of the week, commuters gather on crowded platforms in train stations to ride fantastically congested trains to work and to school. The majority are men dressed in blue, grey, or black suits. Their ranks are interspersed with women in business attire, uniformed high-school and junior-high-school students, and fashionably dressed university students. Forming queues at designated points behind the yellow line, they wait patiently and silently for the train. Some pass the moments before the train arrives by listening to music through headphones, composing emails, or surfing the web on smartphones. Others read pocket novels, magazines, manga, or newspapers carefully folded into fourths in order to negotiate the dense crowd. When the train arrives, it comes to a precise stop so its doors align with the queued commuters. As the doors open, a platform melody commences. Each platform and each station has a distinct melody whose tone, volume, and quickening cadence is calculated to expedite the exchange between embarking and disembarking passengers. Commuters waiting to board know that when the music stops, the doors will close, and they separate into two lines on either side of the train doors, forming a corridor for the arriving passengers to stream onto the platform and into the crowds that are inching down platform escalators and stairs. When the last commuter has alighted, those from the platform surge forward and into the train with those last in the queue entering backward, facing the platform and pushing with their backs into the mass of tightly compressed bodies. As a platform attendant announces through a wireless microphone that the doors will be closing, other platform attendants stand ready to rush forward and nudge a protruding arm, leg, or body inside the train. Finally, the doors are closed, and the repacked train pulls away. From start to finish, the entire operation takes less than thirty seconds.

No words are exchanged among commuters during the process, and no words will be exchanged for the entire ride, leaving the train car in an absolute silence punctuated only by regular service announcements and reminders from the conductor. As soon as one packed train leaves the station, the imminent arrival of the next packed train is announced for the replenished crowd of commuters already queued on the platform.

The first two chapters of this book engage this scene from within the commuter train network's margin of indeterminacy. These chapters prepare the ground for a critique of the ethical quality of the commuter-train collective that I develop in the subsequent chapters. My focus in the first two chapters is on the genesis of the commuter train network, by which I mean the emergence of its underlying schema of operation beyond capacity rather than the history of its technological development. Chapter 1 traces the emergence of operation beyond capacity through a series of phases of rapid urbanization over the course of a century in Japan. I show that in each phase the underlying imperative for train operators involved the question of how to accommodate the rising number of commuters. While the story of urbanization in Japan is a well-traveled history, it is typically told within the framework narrative of technological modernity as a process of progressive rationalization and the corollary mechanistic conditioning of commuter bodies. In contrast, I focus on the way operation beyond capacity evolves by expanding the commuter train network's margin of indeterminacy through a tactical dynamic that I call "finessing the interval." The term reflects the underlying technicity of the commuter collective. It underscores how, with each technological advance and each new tactic to increase operational efficiency that are developed during the phases of rapid urbanization, we see not the tightening of the technological noose around the figurative neck of commuter humanity, but rather a heightened level of openness—an expanded margin of indeterminacy—within the system that elicits increasingly higher degrees of skill and active attention from the commuter. In other words, the more the system suspends "final determination of its own form" under the pressure to operate beyond capacity, the more it relies on the delicate virtuosity of commuters as individuals and groups in order to maintain a collective coherence. The second chapter pursues the genesis of operation beyond capacity through the technicity articulated within the spatiotemporal confines of the commuter-train car. Asking how commuters inhabit the margin of indeterminacy, chapter 2 considers the forms of active attention and the techniques that commuters cultivate in conjunction with the paradoxical mediations of the system. At the same time, it looks at how various forms of media that have become essential features of the commute in recent years—from cell phones to screens to posters—become folded into the spatiotemporal contours of the system's margin of indeterminacy.

Once the genesis of a technical ensemble has been defined, writes Simondon, we can begin to explore the relations between it and other realities.[45] Chapter 3 takes up the question of the technicity of the commuter train network in conjunction with a critical reflection on the kind of intervention the term *technicity* provides into questions of techno-ethics. What I offer is not so much a criticism of the term but rather a critique of the way an agile corporate capitalism has been able to deftly maneuver in recent decades to colonize the conceptual interventions of machine thinking. Specifically, the chapter deals with the development in the late 1980s of a novel, decentralized computer technology inspired by organic systems that allows for administering the margin of indeterminacy as a self-organizing emergent order. I show that, by treating operational irregularities as part of the regular order, the new technology works to transform the underlying organizational schema of the commuter train network from operation beyond capacity to operation without capacity. In so doing, it produces a technological infrastructure that is both resilient to extreme operational events and generative of boundless consumption. The new technology, I argue, comes to embody what has become a general infrastructural paradox: while it realizes a form of infrastructure able to withstand increasingly extreme environmental conditions caused by rampant capitalism, it simultaneously provides a novel schema of operation for modes of extreme capitalism. By tracing the emergence of the new technology in a series of crises of capacity beginning in the late 1960s, my discussion demonstrates that extreme infrastructure has not materialized just in time to help us contend with the inescapable disasters of tomorrow; rather, it is a product of the manufactured crises of the past and the failures of the present.

Turning from matters of technicity to remediation, chapter 4 explores convergences between the internet and the commuter train. The underlying question of the chapter is how the internet thinks the space and time generated in the margin of indeterminacy. This question derives in part from a consideration of the ways in which twentieth-century cinema provided a medium through which to represent the commuter train and the commuter experience—often critically—as expressions of the structural imperatives and contingencies of mass-mediated capitalist society. In asking how the internet thinks the train, I am interested in the ways the internet enables simulations of commuter space and time, and how, in so doing, the internet encourages a shift from knowing and deploying the train in a representational mode to experiencing the margin of indeterminacy differently through performative registers and strategies. Of particular importance in this regard is the way simulation takes its cue from computer gaming. Chapter 4 explores three examples of remediating the train through the internet; in each, computer gaming provides a model for the dynamic, interactive, and performative experience invoked through simulation. At the same time, by transforming commuter space into a kind of game space, the computer-game model produces an invitation to critically engage the experience constituted in the margin of indeterminacy as a place of potential social transformation that begins from the question, *Can the train teach us to care?*

Chapter 5 embarks from within a forty-four minute gap opened within the margin of indeterminacy as a result of a commuter suicide. A commuter suicide, as I show in the chapter, constitutes an extreme event that threatens the network's operational integrity. More importantly, the body on the tracks, I argue, generates a disorder that poses an ethical challenge to the nature of the commuter collective. My concern in the chapter is to understand how the collective attends to that ethical challenge. I show that, on the one hand, the ethical challenge is deferred through a logic of recognition that forecloses acknowledgment of the body on the tracks by reducing it to a mundane and meaningless repetition of salarymen death.[46] The result, I argue, is a collective that is functionally coherent but ethically impaired. On the other hand, the body on the tracks stages a return as a material force demanding acknowledgment. In this context, I turn to a former JR East employee who was tasked with cleaning up after commuter suicides. The chapter finishes, however, with a close reading of the film *Suicide Circle*, in which I argue that the representation of mass commuter suicide brings forward the issue of recognition versus acknowledgment within the context of a complex critique of mass-mediated connectivity.

In the final chapter, the location shifts from Tokyo to a JR West train line just outside Osaka, where a commuter train racing to recover from a ninety-second delay in late April 2005 derailed on a curve near Amagasaki Station, taking the lives of 106 commuters and the train driver. Insofar as the commuter networks of Tokyo and Osaka reflect the distinct character of their respective cities and thus are hardly interchangeable, there are enough similarities between JR West and JR East (in Tokyo) to warrant the accident's inclusion in this book. In particular, JR West is related through its corporate structure and history to JR East. What is more, as the result of a ninety-second delay, the Amagasaki derailment brought forward questions concerning the gap and operation beyond capacity. Although the Amagasaki derailment occurred six years prior to the massive earthquake and tsunami that led to core meltdowns at a nuclear-power plant in Fukushima, it has since become impossible to talk about technological accidents in Japan without referring to this latter catastrophe. It is not difficult to identify a number of parallels between the Amagasaki derailment and the Fukushima nuclear accident in terms of the determinants and effects of massive technological failure in each case. Furthermore, in both cases a technological accident was deemed to be "unthinkable."

My argument in chapter 6 goes beyond such a mode of comparative inquiry to suggest that the Amagasaki derailment asks us to think about the problem of risk and about technological accident in general as a manifestation of a problematic relationship with machines. What we discover in the wake of the Amagasaki derailment is not a community of commuters longing for a return to a simple existence free of complex technological systems like train networks, power plants, and airplanes, but rather a community that takes it upon itself to engage seriously with the problems of complicated technological ensembles and to think about the possibility of a different kind of relationship with machines engendered through a remediated structure of institutional and social governance. In this context, chapter 6 draws attention to how, for the commuter community and the victims of the Amagasaki derailment, the possibility of developing a different kind of relationship with machines and technological risk involved a reexamination of the gap—the ninety-second delay behind the accident that was at the center of the subsequent investigation and controversy. Whereas the train company attempted to attribute the gap to human error on the part of the train driver, the community came to understand the gap as an effect of a *daiya* without a sustainable margin of indeterminacy. This is what I call "thinking with the gap." Consequently, while the train company endeavored to close the gap by relegating its significance to a matter of technical modification, in thinking with the gap the community sought to hold the gap open as a problem space in which to engage with and reflect on the values that had allowed for the emergence of a technological ensemble without a sustainable margin of indeterminacy. As such, the gap becomes a space though which to reconceptualize not only the nature of institutional trust but also the capacity of technological ensembles to be trustworthy.

Privatization, the Bubble Economy, and Neoliberalism

Tokyo has had a thriving commuter train system since the early decades of the twentieth century, especially since the reconstruction of the city following the Great Kanto earthquake of 1923. While I explore some of the key developments in the early years of the system in chapters 1 and 2, the central time frame in this book is the postwar years to the present. Within this period, the privatization and breakup of Japanese National Railways (JNR) in 1987 and the collapse of the nation's bubble economy in the early 1990s figure as paramount events in terms of their impact on the technological and social organization of the commuter train networks in both Tokyo and Osaka. The privatization and breakup of JNR marked the end of an era. Formed just over eight decades earlier in the Railroad Nationalization Act, JNR was an institution of considerable economic and social significance whose iconic and material inseparability from the nation was captured in the idiom "JNR is the legs of the nation" (*kokutetsu wa kokumin no ashi*).[47] It was thus thought to be indissoluble without risking the national economy. Consequently, privatization of JNR in the spring of 1987 was the final act in a long and bitter struggle between the Ministry of Transportation, JNR management, and railroad labor unions that had begun in the 1960s and 1970s with the government's series of failed campaigns to rationalize railway labor.[48] Privatization broke JNR into six regionally based passenger railways that included JR East in Kanto around Tokyo and JR West in Kansai around Osaka, Japan's second-largest city.[49] Although these events occurred nearly two decades before I began the research for this book, the tensions generated around that major historical change remain palpable, especially around conflicts such as the Amagasaki train accident (chapter 6).

Similar to the privatization of JNR, the collapse of the nation's bubble economy in the early 1990s transformed the social and cultural landscape, especially in Japan's urban centers.[50] Japan's bubble economy was the result of complex financial finessing that spurred stock-market speculation and overlending from banks working with inflated land values. It extended roughly from the mid-1980s to early 1990, when the nation's stock market began to collapse. The collapse sent Japanese society into a tailspin. The nation saw one prime minister after another unable to fulfill his tenure, while once-stable large companies rushed to restructure their workers, resulting in an unprecedented number of layoffs. The collapse was followed by years of economic recession that came to be known as "lost decades." During this time, the government moved to deregulate labor and slash social funding.

In more recent years, a number of scholarly publications have linked the above moments of historical transformation and unrest with the rise of neoliberalism in Japan. In particular, these scholarly works draw attention to the way neoliberal-style reforms of the political economy in Japan have led to increasingly precarious structures of employment,[51] the valorization of independence and self-responsibility,[52] shifts in education policy,[53] and the rise of the enterprise-society ideal.[54] The subject of neoliberalism also enters my argument in this book, particularly in my discussion of the development of new *daiya* technology (chapter 3) and the Amagasaki accident (chapter 6). However, I treat neoliberalism as a perversion of machine theory rather than as an effect of political economic theory and government policy. In this regard, my argument dovetails with Melinda Cooper's identification of a specific post-1970s iteration of neoliberalism that emerges with epistemological shifts at the intersection of theoretical physics, life sciences, and computer sciences.[55]

Post-1970s neoliberalism is economic theory under the influence of nonlinear complex-metastable-systems theory. Or, rather, it is the co-opting of the principle of metastability from theories of emergence as a means of providing quasi-scientific rationale for the economic exploitation of the protean qualities of organic life—what Cooper calls capturing "life as surplus." Whereas Cooper focuses on the conversion of the principle of metastability in economic theory into socioeconomic precarity, I am interested in the materialization of emergence in extreme infrastructure and its appropriation toward realizing a new form of extreme capitalism that exploits the dynamic qualities of collective life. Neoliberalism, in my argument, is thus not just the expression of government or corporate economic policy: it is the result of subjugating to economic form the attempts in a number of related academic fields to overcome material limits by thinking with technology.

In the Shadow of Fukushima

No event in recent decades has raised more concern in Japan regarding technology than the reactor meltdown at the Tokyo Denryoku (TEPCO) Daiichi nuclear-power plant in Fukushima following the earthquake and tsunami of March 2011. In the initial years following the disaster, crowds of tens of thousands filled the streets and squares in front of the Japanese prime minister's residence and government buildings demanding that the government hold TEPCO accountable for the meltdown and scrap the nation's nuclear-energy program. Despite TEPCO's continuing struggle to clean up and contain the radioactive leakage from the event, the urgency behind those demands seems to have peaked. Antinuclear demonstrations in Tokyo continue, but they have become routinized events that draw a fraction of the participants they did in the past and are unable to produce enough energy to effect change.

The disasters in northeast Japan and the Fukushima reactor meltdown lend exigency to the argument in this book for a reconceptualization of our relationship with technology through the use of machine theory. Overwhelmingly, the reactor meltdown at Fukushima and the failure of disaster infrastructure in northeast Japan have been dealt with in terms of an ethical problem of a political and economic nature, with scholars pointing to multiple levels of collusion between the nuclear industry, construction companies, and government.[56] Machine theory generates a different kind of intervention. It asks us to conceptualize a relational ethics from within the margin of indeterminacy of a collective rather than as a rationally conceived formula meant to provide ethical order to a material world. Thus, machine theory elicits a techno-ethics that is highly particular in its formulation, demanding attention to the specific kinds of relationality enabled in a technology. Machine theory demands that we attend to technology in terms of its technicity—the quality of its collective. Following machine theory, a nuclear-power plant would not be thinkable as a solution for a collective's energy needs.

Although the ongoing nuclear crisis at Fukushima only enters this book explicitly in the last chapter, it is an implicit impetus behind the entirety of my argument. Since March 2011, the Fukushima meltdown has been a key part of the conceptual milieu in which the ideas and sense of urgency for this book were formed. In this context, this book is an effort to think with a technical ensemble from a novel direction, in order to mobilize critical intervention into the kind of thinking that has given rise to such things as nuclear-power plants, and to reimagine the possibilities of collective life.

NOTES

1. See page 35 of www.mlit.go.jp/common/001179760.pdf.
2. http://kikakurui.com/e/E7106–2011–01.html.

    Commuter train car capacity is determined by the Japanese Industrial Standards Committee. Based on the determination that the average commuter weighs between 55 to 60 kilograms, the committee stipulates that each commuter requires 430 millimeters of seat space and 0.3 meters of floor space, meaning three commuters to every square meter of space. According to Ramon Brasser, a researcher at the Tokyo Institute of Technology's Department of Earth-Life Science, if we begin from a calculation that the typical person (55 to 60 kg) is 0.4 meters in width and 0.24 meters in depth, then the absolute minimum space required for one body is 0.096 square meters ($0.4 \times 0.24 = 0.096$). Accordingly, ten people can fit into one square meter of space only if they are stacked liked sardines in a can (www.elsi.jp/en/blog/2015/11/blog1126.html, accessed March 26, 2017). Certain points in Tokyo's commuter train network, such as between Nakano and Shinjuku Stations on the Chūō Line, realize such intense sardine-like congestion conditions during the morning rush hours.
3. I borrow this example from Mito, *Teikoku hassha*.
4. Tomii, *Resshya daiya no himitsu*.
5. Mito, *Teikoku Hassha*; Tomii, *Resshya daiya no himitsu*. The setting was chosen more out of concern over the logistics of gathering hundreds of railroad employees for an extended time rather than because of any potential for indulgence. The process involved a number of phases, including at one point coordinating lengthy negotiations over rights-of-passage among managers from different train lines in each area of the country.
6. Tomii, *Resshya daiya*.
7. Eguchi, "Ressha daiya wa ikimono," 103. My translation. (Throughout this book, unless otherwise noted, all translations are my own.)
8. Simondon, *On the Mode of Existence of Technical Objects*.
9. Jansen and Vellema, "What Is Technography?"; Vannini, Hodson, and Vannini, "Toward a Technography of Everyday Life."
10. I am referring specifically here to approaches for the exploration of technology. See for example the notion of "ontological experiment" in Jensen and Morita, "Infrastructures as Ontological Experiments," as well as the notion of "practical ontology" in Gad, Jensen, and Winthereik, "Practical Ontology: Worlds in STS and Anthropology." In both of these works, Andrew Pickering's call for a "performative idiom of knowledge" is an explicit inspiration; see Pickering, *The Mangle of Practice*.

11. Even as recently as the early 1990s, the sociocultural anthropologist Bryan Pfaffenberger argued in his review of anthropological approaches to technology that anthropology had yet to take technology seriously as a legitimate topic of inquiry. Pfaffenberger, "Social Anthropology of Technology."

12. Marc Augé's work on the non-places of technological mediation and the Paris Métro stands out as an exception; see Augé, *In the Metro*, and Augé, *Non-Places: Introduction to an Anthropology of Supermodernity*. Despite working to bring the subject of large technical ensembles into the terrain of anthropological inquiry, Augé did not develop a corresponding theory of technology. Although innovative in many respects, his work stays within the confines of ethnographic theory.

13. Such concerns were most forcefully and famously articulated by key thinkers associated with the Frankfurt School, whose work on technology veered toward technological determinism. The Frankfurt School is the term given to an intellectually diverse group of scholars associated, sometimes somewhat tangentially, with the Frankfurt Institute for Social Research, founded in Germany in the interwar period. Among its central figures, the most outspoken in their concern with technology were Theodor W. Adorno and Max Horkheimer. Although Walter Benjamin and Georg Simmel (who are also associated with the Frankfurt School) diverged from Adorno and Horkheimer's position, they tended to treat questions of technology mainly within their larger preoccupation with developing a theory of modernity.

14. The work of Bruno Latour, particularly his development of Actor Network Theory (ANT), was especially responsible for opening anthropology up to large technical ensembles.

15. Simondon, *On the Mode of Existence of Technical Objects*, 157.

16. Hayles, *How We Became Posthuman*; Johnston, *The Allure of Machinic Life*; Martin, "The Organizational Complex"; Mindell, *Between Human and Machine*; Pickering, *The Cybernetic Brain*.

17. Halpern, *Beautiful Data*.

18. Wiener, *The Human Use of Human Beings*. Wiener's reductive approach posits technological systems and social systems as comparable expressions of a logical, informational patterning. His theory has been widely criticized for leading cybernetics into its failed attempts at building artificial intelligence in the 1960s; see Brooks, "Intelligence without Representation"; Brooks, "Intelligence without Reason"; Johnston, *The Allure of Machinic Life*.

19. Parikka, *Insect Media*, 141.

20. See Thomas LaMarre's critique of the cyborg notion in LaMarre, "Afterword: Humans and Machines," in *Gilbert Simondon and the Philosophy of the Transindividual*.

21. Latour, *We Have Never Been Modern*; Latour, *Reassembling the Social*.

22. Haraway, *When Species Meet*.
23. Gad, Jensen, and Winthereik, "Practical Ontology"; Morita, "The Ethnographic Machine."
24. Combes, *Gilbert Simondon and the Philosophy of the Transindividual*, 71.
25. Simondon, *On the Mode of Existence of Technical Objects*.
26. Fujii, "Intimate Alienation"; Hashimoto and Kuriyama, *Kindai nihon ni okeru tetsudō to jikan ishiki*; Kuriyama and Hashimoto, *Chikoku no tanjō*; Tanaka, *New Times in Modern Japan*; Noda, Katsumasa, and Eichi, *Nihon no tetsudō*; Sawa, "Nihon no tetsudō koto hajime."
27. Schivelbusch, *The Railway Journey*.
28. Ibid., 74–75. Schivelbusch cites Simmel in the following manner: Georg Simmel, *Soziologie* (Leipzig, 1908), pp. 650–1.
29. For literature on the train in Japanese, see Nakamura, *Ressha seigyo*; Nihon kokuyū tetsudō sōsaishitsu shūshika, *Nihon kokuyū tetsudō hyakunenshi*; Noda, Katsumasa, Eichi, *Nihon no tetsudō*; Sawa, "Nihon no tetsudō koto hajime"; Watanabe and Tamura, *Ryojō 100-nen*; Kuriyama and Hashimoto, *Chikoku no tanjō*.
30. Simmel, "The Metropolis and Mental Life."
31. LaMarre, *The Anime Machine*, xvii.
32. Lévi-Strauss, *Totemism*.
33. LaMarre, *The Anime Machine*, xxvii.
34. Ibid., 301.
35. Mackenzie, *Transductions*, 53.
36. Ibid.
37. Sarti, Montanari, and Galofaro, *Morphogenesis and Individuation*.
38. Simondon's formulation of the term *information* departs considerably from his understanding of the term *cybernetics* in terms of a quantifiable signal, probability, and entropy. As Thomas LaMarre points out in the introduction to Muriel Combes's work on Simondon, if cybernetics gives us information theory, Simondon produces a theory of information (Combes, *Gilbert Simondon and the Philosophy of the Transindividual*, xv). In the latter, information is a nonquantifiable material force, or what Jussi Parikka parses as "the intensive process of change at the border of different magnitudes" (Parikka, *Insect Media*, 142).
39. For an excellent distillation of the challenges posed by the anthropology of infrastructure, see Larkin, "Politics and Poetics of Infrastructure."
40. The term "non-places" is from Marc Augé's work of the same name, *Non-Places*.
41. My use of the term "indeterminate publics" is in reference to Mimi Sheller's methodological problematization of the "mobile public" of transport and communications infrastructure; Sheller, "Mobile Publics."
42. Anthropologists have handled this methodological dilemma in creative ways. Most notably, Bruno Latour's Actor Network Theory (ANT) endeavors to capture the complexity of the collective by mapping its constitutive

human and nonhuman relations; Latour, *Reassembling the Social*. While Latour's approach has proven generative, it has been criticized for its failure to engage the ethical quandaries of these relations; see Fortun, "From Latour to Late Industrialism." By contrast, others have brought forward ethical considerations by foregrounding the political, legal, and economic tensions manifest in infrastructural processes; see Chu, "When Infrastructures Attack"; Harvey and Knox, *Roads*; Anand, "Pressure"; Appel, "Walls and White Elephants."

43. Simondon, *On the Mode of Existence of Technical Objects*, 157.
44. Ibid.
45. Simondon, *On the Mode of Existence of Technical Objects*, 20.
46. *Salaryman* is a broad term incorporating normative notions of gender, class, and race that appears early in Japan's history of modernization but has seen significant fluctuations in its currency in recent decades. Basically, it refers to a salaried, full-time, white-collar, male Japanese worker.
47. JNR was initially designated the Japanese Government Railways. It was renamed JNR after World War II.
48. Weathers, "Reconstruction of Labor-Management Relations in Japan's National Railways"; Kasai, *Japanese National Railways*.
49. The Japan Railways (JR) Group is an organization comprising six passenger-rail companies, each operating within a designated geographical region (JR Hokkaido, JR Central, JR East, JR West, JR Shikoku, JR Kyushu), a freight company (JR Freight), a research organization (RTRI), and an information-systems company (JR System). The group includes a number of subsidiary companies as well, such as the East Japan Marketing and Communications Company (JEKI), which handles advertising within the JR East Kanto Network.
50. I conducted the initial fieldwork for this book between spring 2004 and winter 2006. During that time, I lived on the western side of the Greater Tokyo Metropolitan Area in the city suburb of Higashi-Koganei, which is served by the Chūō Line. I then returned to Tokyo, where I remained from late 2006 until the summer of 2008, accompanied by my wife, who was engaged in her own anthropological fieldwork. Since 2010 I have had the opportunity to return to Tokyo almost every summer in order to conduct follow-up research and interviews.
51. Allison, *Precarious Japan*.
52. Alexy, "Intimate Dependence and Its Risks in Neoliberal Japan"; Lukacs, "Dreamwork."
53. Arai, *The Strange Child*.
54. Yoda, "A Roadmap to Millennial Japan."
55. Cooper, *Life as Surplus*.
56. Kainuma, *Fukushima ron*.

ONE

Finessing the Interval

You are packed into the train so tight that you feel as if your internal organs are going to be crushed. By the time I arrive at work, I'm exhausted and too tired to do anything. I would do anything not to have to ride the packed train but there is no choice [*shōga nai*].

TOKYO COMMUTER (A PARALEGAL AND LAW STUDENT)

"There is no choice": one hears this phrase often from commuters in Tokyo regarding the packed morning commuter train. Insofar as this phrase seems to offer a succinct and compelling explanation for how and why commuters endure, day after day, the fantastic compression of the packed train, it also has the unfortunate effect of reducing the packed train to a spectacle of compliance. As such, the packed train becomes a mere trope, a metaphor for totalizing forces that lie elsewhere, outside the train, in either the historical processes of technological modernity or the unique and immutable relations of Japanese culture. While the former renders the packed train a discursive effect of the forever-escalating processes of rationalization under industrial modernity, the latter hints at a particular, culturally ingrained, and pathological disposition toward the "authoritarian personality."[1]

How might we understand the packed train in a nontotalizing way, as something other than an expression of capitulation and compliance? Put differently, how might we grasp the packed commuter train as an ongoing process of collective making involving the coemergence of humans and machines? A similar question inspired Gilbert Simondon's efforts throughout the 1950s and 1960s

29

to remediate the conceptual framework for understanding humanity's relationship with technology.[2] In contrast to dominant discourses of the time, which tended to focus on the history of technological development and the social impact of technological objects, Simondon proposed understanding technological elements, machines, and ensembles in more evolutionary terms, from the perspective of their genesis within a particular milieu of relations.[3] In shifting the approach from history to genesis, Simondon reframed the overarching question from how is a technology *formed* through design—and what is its corollary determining effect on humans—to how does a technology *take form* through the work of human innovation in conjunction with the conflicts and relationships specific to its milieu. Design, according to the latter formulation, can be understood in terms of what Félix Guattari calls a heteropoietic process, in which thinking is elicited from material relations rather than being something born of the human faculty for abstract reasoning and imposed on the world.[4] The critical question then becomes, To what extent does the technology remain "*in formation*," which is to say processual or structurally underdetermined such that its performance is able to vary in accordance with information received from its operating environment?[5] This tension between structure and process, routine and contingency animates an organism's or technology's margin of indeterminacy.

Simondon's thinking thus places importance on the degree to which a technical ensemble's margin of indeterminacy permits it to enfold an expanded network of relations—an expanded collective entanglement.[6] Living organisms, in Simondon's thinking, have a high margin of indeterminacy, which means that they can continue to individuate, or form associated milieus within their environment.[7] By contrast, a technical element, machine, or ensemble is understood as having an established or at least far more limited margin of indeterminacy.[8] The extent to which a technology is able to realize a coherent unity of relations while maintaining a margin of indeterminacy is its "degree of concretization," which Simondon parsed as its technicity.[9] Thus technicity is not in a technological object itself but within the *quality* of its constitutive and emergent relations. Accordingly, it draws our attention to the specific nature of collectivity that forms in connection with a technical element, machine, or ensemble.

Deleuze and Guattari take up Simondon's genetic approach to technology with the notion of a "machinic phylum." A genetic approach offers a novel methodological intervention in encouraging us to ask, as Deleuze and Guattari did about the body, not what a technology *is* but both what a technology *can do* and what the limits and possibilities of its collective are.[10] These are the questions I take up in this chapter when exploring the genesis of Tokyo's commuter train network over the course of a century by examining a series of formative phases of rapid urbanization. In tracing these phases, I underscore how the increasing pressure for operation beyond capacity involves the evolution of a tactical dynamic I call *finessing the interval* within the commuter train network's margin of indeterminacy. To finesse something is to make it work when, logically speaking, it should not. Finesse is about pulling something off against all odds. Invoking terms like *flair*, *panache*, or *élan*, finesse bespeaks a method irreducible to skill, expertise, or systematicity. Finesse transcends the logic of rational methods whereby cause and effect can be situated as calculable corollaries; it involves instead qualities like instinct, affect, and feeling—qualities that are embodied, sensual, and informed by the precarious order of contextual relations. Insofar as finesse bespeaks a human capacity for delicately orchestrating an event or relations, it is also never far from the notion of *machination*, a term that refuses a simple division between humans and technology. Denoting a kind of trickery effected by virtue of device or contrivance—also terms that refuse simple ontological categorization—machination invokes the notion of a relation between the human and the technological. More importantly, it suggests a relation that transpires as a kind of dialogue in the mode of technicity between provisionally stable processes rather than established and fixed ontologies. Humans and technological ensembles enter into a dialogue only by virtue of a certain openness, an unfinished quality of the ongoing processes that animate both.

In tracking the genesis of operation beyond capacity in Tokyo's commuter train network, I am not suggesting that residents of cities throughout the world need to learn to compress themselves into impossibly packed trains day after day, or even that they should learn to queue in an orderly fashion on the station platform. Tokyo's commuter train network and operation beyond capacity are not models of techno-social organization to be emulated. Rather, they are a collective condition that is good to think with. My argument is that in order to grasp the technicity of the commuter train network we must be able to think with it and reimagine our relationship with technology. In this regard, I do not contend that the notion of capitulation and compliance is wrong, but that it is far too simple in being premised on an underlying adversarial relationship between the human and the technological. Machines do not impose determinations; they elicit relationalities.

Similarly, technological systems do not work because they are precise and constraining—or rather, they do not work *well* when they are precise and constraining. They work when they engender collectivity—that, is when they allow for a kind of co-constituting and mutually beneficial dialogue between bodies and the environment. In this co-constituting labor, collectivity emerges. Operation beyond capacity in Tokyo's commuter train network is exemplary in this regard because it simply would not work if commuters merely capitulated to and complied with the system. The network functions only because of the constant, attentive labor of commuters to adjust to and become with the collective. In short, I am not presenting the genesis of Tokyo's commuter train network as a normative paradigm for a better technological world. I offer it instead as an analogy to think with and thus, hopefully, think more complexly about the limits and possibilities of our relationships with and within technological environments.

Accommodating the Masses

Beginning in the early decades of the twentieth century, commuter demand in Japan persistently increased. There is no shortage of detailed records regarding this persistent increase;[11] there are also numerous scholarly works historically contextualizing it. All of these tell a more or less similar story, centering on three significant phases of development. The first phase was during the rapid urbanization between 1914 and 1918, when the need to supply allied forces during World War I spurred an industrial boom.[12] The second phase began following the Great Kanto earthquake of 1923, which caused widespread devastation and fire throughout much of Tokyo. Prior to the earthquake, Osaka (in Kansai) rather than Tokyo was the nation's center of industry and rail development. In that region, rail industry was dominated by private train companies that adopted a commuter-consumer paradigm, developed overseas, of building department stores around terminal stations and expanding lines to reach exclusive resort and recreation towns.[13] Following the Great Kanto earthquake of 1923, Japan's national railways took the lead. Reconstruction transformed Tokyo into a modern commuter city with much of the population living in the suburbs and commuting into the city center for work, entertainment, and shopping, placing increased demand on the city's rail networks.[14] Finally, the third phase occurred in the initial decades of the post–World War II era. The program of intense economic growth that was orchestrated under centralized government planning during this period created un-precedented demand for commodity and commuter transport.[15]

In each of these phases, train companies faced the dilemma that the demand for transport exceeded the rational limits of the infrastructure's capacity. This dilemma could not be resolved simply by adding rolling stock (more trains), enhancing system performance, or providing further commuter training. For one thing, access to rolling stock was for a long time severely limited. More importantly, even if train companies realized perfectly punctual operations and managed to discipline the commuter population to march like automatons in precise rows into train cars—the dream of fascist nations—the result would have been inadequate to the demand for transport. The system had to be capable of both handling commuter crowds that could barely fit on station platforms and transporting numbers far beyond the system's capacity. In other words, it is not simply that rationalization alone would not have been enough: it would not have worked at all. Such circumstances demanded not a structurally perfect system of absolute precision and compliance but rather a tactic of finessing the interplay between human and machine. What I call "finessing the interval" refers to precisely such a tactic, which took shape in the interwar period as part of a technique for "accommodating passengers."

In a fascinating and exhaustive history of the development of Japan's railroad system, the Japanese economist and infrastructure historian Mito Yuko describes the emergence of the technique of "accommodating passengers" in order to meet the ever increasing commuter demand.[16] The phrase derives from the Japanese idiom "accommodating customers" (kyaku o sabaku), which refers to the way a restaurant or retailer manages to serve far more customers than the given infrastructure and number of personnel would otherwise permit. Mito offers the example of the typical ramen restaurant in the city with limited counter space that nonetheless accommodates a huge lunchtime crowd. How the ramen shop manages to serve beyond its structural capacity cannot be understood as merely the result of an efficient use of staff, material setting, and customer cooperation. The technique instead emphasizes the interplay between the three. The space and time of interplay is treated as dynamic—as a dimension with a certain parameter for actions, behaviors, and responses—rather than as a setting with a specific operational script. Both the customers and the operators of the ramen shop must remain keenly attuned to the dimension of interplay and ready to adapt to the modulating conditions. Attunement materializes as an embodied and distributed attention interwoven with the setting rather than consciously directed. Attunement involves "sensing the air" (*kūki o yomu*). It is about the pressure and heat of the air between bodies and things, the texture of smells and sounds that constitute a collective space. Attunement transpires as the capacity to feel *with* the intensities of one's surroundings, which manifest in the materializations of difference that ripple across the fabric of heterogeneous realities woven into an atmospheric flow. The customer in the crowded ramen shop, even while hunched over a bowl and absorbed in the deeply satisfying process of slurping down hot noodles and broth, remains tuned in to these ambient shifts in intensity.

In much the same way, the technique of accommodating commuters for operation beyond capacity in the commuter train network can't be reduced to the efficient performance of personnel and machines or the training of commuters to comply with a strict protocol. Emphasis falls rather on the interplay between humans and machines and the capacity therein to finesse an optimal dynamic in order to perform beyond expected limits. Within the commuter train network, that dimension of interplay is the system's margin of indeterminacy. As with the ramen shop, attunement and adjustment to the constantly shifting conditions of the dimension of interplay from commuters and personnel, rather than a strict adherence to a predetermined order, is critical. Such attunement transpires on neither a conscious nor subconscious level. It is the effect, rather, of a collective and distributed labor of remaining tuned in to a persistent background connectivity that is registered and relayed corporeally in the actions, behaviors, and responses among commuters. The capacity for such collective attunement results not from an inherent cultural disposition but rather from a relation between commuters and the commuter train network that has evolved over the course of a century in conjunction with the schema of operation of finessing the interval. I will return to the labor of collective attunement among commuters in the next chapter; in the meantime, I want to track how finessing the interval emerged as an infrastructural schema of operation in Tokyo's commuter train network.

As Mito explains, the years around the First World War were critically formative in the development of this schema of operation. Demand for transport increased sharply during this time, nearly tripling on main lines, which insisted that train companies increase their rolling stock.[17] However, access to new rolling stock was extremely limited. Japan had not yet developed a robust manufacturing industry of its own, and imports from the United States and Europe were curtailed by the international conflict.[18] Train companies thus had to increase the traffic density (number of trains per hour) without the benefit of new trains or tracks.[19] The solution developed was twofold: (1) decrease the time required for trains to complete a run, and (2) decrease the turnaround time at the terminal station so as to put the train back into circulation more quickly.[20] This involved several "speed-up" strategies: First, running time between stations was decreased by increasing the overall speed of the train. Second, turnaround time at terminal stations was shortened in part by training cleaning crews to be on standby even before trains arrived at terminal stations. Finally, and most important, dwell time (the stopping time at stations) was decreased from minutes to seconds by expediting the boarding and alighting of passengers.

As a result, between 1914 and 1918 the dwell time at major stations in Tokyo was reduced from around two minutes to one minute or less. At midsize stations, it was reduced to thirty seconds.[21] The strategies developed during this period carried over into the next phase of post-1923 Tokyo. While the period witnessed a number of technological advancements, including the electrification of major commuter-train lines, it also saw dwell time at stations throughout Tokyo during peak hours of congestion decrease to a standard twenty seconds.[22] By 1924, only one year after the Great Kanto earthquake, during rush hours Tokyo's Yamanote Line made a complete circuit in sixty-two minutes and forty seconds (compared with fifty-nine minutes today) while trains on the Chūō Line between Nakano and Tokyo Stations ran with a three-minute headway. By 1925, trains on the Tōkaidō Line between Tokyo and Shinagawa Stations operated with a two-and-a-half-minute headway.[23] With trains spaced so closely, there was no longer a need for stations to offer waiting rooms, and station architectural design became focused instead on facilitating the continuous flow of commuter bodies through the structure.

Decreasing dwell time at stations was never merely a technological matter. It was a process in the evolution of a technicity in which commuter bodies and commuter-train apparatuses entered into a dialogue around finessing the interval for operation beyond capacity. As such, it was a process of intensifying dynamic that involved greater effort from commuters, eliciting from them an increasingly aggressive attunement to the modulating ambiences of the system's environment. Part of what drove this intensifying dynamic was a basic paradox: the more the system was asked to operate beyond capacity, the more dwell time was needed to accommodate crowded platforms, and yet the more commuter-train operators needed to decrease dwell time in order to accommodate the commuter population.

Although the logic is straightforward, it did not stop researchers from trying to test its limits in an experiment conducted in Tabata Station, Tokyo, on June 5, 1926. Inspired, perhaps, by the brief rise in Japan of a Taylorist-style "efficiency movement" at the time, the experiment was organized by the Railway Training Institute and involved 295 young students, who were gathered on the platform and asked to board and detrain in various numbers and conditions.[24] The train car used in the experiment had a specified capacity of 108 passengers, but was judged able to accommodate at least three times that number, around 320 passengers.[25] The result of the experiment was as follows:

Boarding Times

Empty train car, 50 passengers	5 seconds
Additional 50 passengers	4.5 seconds (total of 100 passengers)
Additional 40 passengers	6 seconds
Additional 60 passengers	10 seconds (total of 200 passengers)
Additional 30 passengers	7 seconds
Additional 15 passengers	7 seconds
Additional 35 passengers	30 seconds (total of 280 passengers)

The last two passengers required 2 seconds.

Combined detraining and boarding time for a train with 200 passengers

100 passengers detrain, 50 board	16 seconds
50 passengers detrain, 100 passengers board	15 seconds
100 passengers detrain, 60 passengers board	15 seconds

The results of the experiment are hardly surprising, demonstrating unequivocally that the more congested conditions are, the more time is required for commuters to board the packed train. Nevertheless, the experiment is important for understanding the genesis of the packed train, since it foreshadows what I explore below as a defining paradox in the process whereby the commuter collective developed in conjunction with the rapid urbanization accompanying Japan's postwar economic recovery.

Very early in this process of postwar economic recovery, it became necessary to further decrease dwell times at stations in order to accommodate the surge in commuter demand, yet dwell time could not be further reduced without encountering the impasse demonstrated in the experiment. The system of finessing the interval had reached a threshold. That threshold, as we will see, was overcome in the early years of the postwar through a combination of technological innovations that allowed for opening yet another interval—an interval within an interval—by means of a dynamic system of recursive temporal debt and recovery in the space and time between stations. What is especially peculiar about this dynamic is that it not only produces a high degree of precarity around the margin of indeterminacy, but also remains entirely informal, meaning that it develops as a mode of collective operation that is only tacitly scripted. This informal dynamic, which I identify below as the dynamic of *yoyū*, comes to define the contemporary technicity of the commuter train network.

Creating *Yoyū* within the High-Capacity/High-Density Network

The first decades after World War II in Japan were marked by a steadfast belief, among the population and its leaders, in the progressive power of "science and technology."[26] Within this period, the mid-1950s saw an unprecedented surge in urban overcrowding as a result of an economic boom precipitated by a combination of political, social, and technological factors that were greatly informed by Japan's relationship with the United States. Having lost China to Communism, the United States was determined, by the late 1940s, to consolidate its position in Japan by promoting quick economic recovery. To this end it initiated its infamous "reverse course" tactic, which involved suppressing Japan's burgeoning postwar labor movement while facilitating mass production through the introduction of advanced automation technology and the opening of the US market to Japanese imports. The real economic boom in Japan began, however, with the opportunity to supply the United States with military equipment and products for its war in Korea.[27] As a result of the production surge from the war, by 1956 Japan's Economic Planning Agency was ready to declare the end of postwar economic recovery and forecast the onset of a period of "high economic growth."[28]

Lured by the promise of employment, Japan's inhabitants, especially fresh university graduates, poured into the cities in unprecedented numbers in the early 1950s. The enormous influx of population produced a phenomenon of urban crowding that became a topic of social critique and concern.[29] During this time, long rows of young commuters filling station platforms, bottlenecked at ticket gates, and crammed

FIGURE 1.1. Rush hour at Shinjuku Station, 1964. It is 8:40 a.m., and the congestion continues.

into train cars became a regular part of the daily commute. This intense daily congestion and the commuting experience it engendered became commonly known as "commuter hell" (*tsūkin jigoku*).

The incredible level of commuter congestion marked a limit for prewar strategies of accommodating commuters, a limit that could be resolved only through new tactics. Neither dwell time at stations nor headway between trains could be reduced any further. As with the strategy of accommodating commuters, the solution that emerged involved a combination of technological, political, and social factors, beginning with the introduction in 1957 of a high-performance commuter train (*101 kei shinseinō densha*).[30] A product of Japan's railroad industry, the new train ushered in a system of high-traffic density and high commuter capacity (high density/high capacity) that remains the basis of Tokyo's commuter train network. The new train exploited a redesigned propulsion system to reduce vibration, a more powerful motor, a combination air/motor braking system, and an air suspension system that hushed the sounds and smoothed the jolts of the apparatus so as to immerse commuters in a new degree of silence. It also incorporated a light construction design developed by aeronautics engineers who had previously worked on Japan's military aircraft but found themselves unemployed with the nation's ostensible commitment to peace in the postwar era.[31] Most significant among these developments was that the combination of these factors lent the train a vastly improved ability for rapid acceleration when departing a station and equally rapid deceleration when entering a station. This would prove critical for finessing the interval even further for operation beyond capacity.[32]

Along with the development of high-performance trains, the early postwar era also saw a concerted effort among train companies to reorganize traffic-control operations along a centralized model of command and control. Beginning in the late 1950s, JNR as well as private railroad companies began introducing Centralized Traffic Control (CTC) technology borrowed from the United States, a topic I will address more fully in the third chapter.[33] With the centralized system, a dispatcher in the system's central control room monitors the progress of train traffic on a large schematic of the system and issues orders to trains and stations. Depending on the level of automation integrated within the system, the dispatcher can remotely operate switches and signals, routing a train to the appropriate platform at a given station. In the United States, CTC technology was introduced to increase transport volume by rationalizing operations. In postwar Japan, by contrast, transport volume was already at or beyond the maximum.[34] The technology was thus adopted as a means to redistribute and reduce labor by centralizing, simplifying, and automating many of the traffic-control functions that had been the responsibility of individual stations.

The work of centralization extended over a number of decades; some JNR lines didn't even receive the CTC technology until the late 1970s. As a project driven by an impulse for totalizing representation, centralization was never fully completed. Even where it was fully completed, it was never entirely successful. Main train lines within Tokyo's commuter train network, such as the Chūō and Yamanote Lines, remained far too complex, generating more information than could be handled in a timely manner by a single central command room. As a result, large stations on such main lines shared the traffic-control burden to form a hybrid centralized/noncentralized command schema. Centralization thus remained an ideal rather than a reality.

More than promoting centralization, the high-performance train provided for a new degree of finessing the interval in the early postwar decades by allowing for the creation of an informal dynamic of recursive temporal debt and recovery in the intervals between stations. System planners realized that the high level of congestion made it impossible for drivers to conform to a strictly scripted traffic pattern. It was understood that in order to accommodate the congestion and pack trains beyond capacity, drivers would need to deviate from the planned traffic diagram by extending the dwell time at stations beyond the allotted time. They were expected to then exploit the improved acceleration and deceleration of the new trains in order to recover the time in the short interval between stations in a practice that in railroad parlance was known as "recovery driving" (*kaifuku unten*). Recovery driving became an integral, albeit informal, part of operation beyond capacity in the postwar period. In opening a temporal gap in the interval between stations, recovery driving creates a dimension of *yoyū*. Composed of the character for "surplus, excess, remainder" (*yo/amari*) and the character for "abundant, rich, fertile" (*yū*), *yoyū* suggests a process of overcoming limits by means of creating a bit of extra space or time where there was none before. As the *Kōjien* dictionary (the Japanese equivalent of the *Oxford English Dictionary*) states, *yoyū* denotes producing an excess of what is required (*hitsuyō na bun no hoka ni amari no aru koto*).[35] At the same time, the notion of time and space that is referred to as *yoyū* encompasses human emotional and psychological limits. Thus, for example, when one uses *yoyū* in the context of having room in one's heart—that is, having compassion for someone or something (*kokoro no yoyū*)—it connotes that one has overcome certain physiological and psychological barriers to create an emotional space in one's psyche and life for someone or something else. Evoking, to a certain extent, the sense of creating something from nothing, *yoyū* can be understood as a spatiotemporal alchemy relying not on external magical forces but rather on the finessing of an interval at work in a relationship, whether human, machine, or both.

The term *yoyū* is sometimes translated as "leeway." For the sake of the technological and social specificity I have laid out here, however, I will continue to use the Japanese term *yoyū* throughout this book. As a concept and practice *yoyū* is irreducible to a system of rationalization or to any technological, social, or human factor. It is an operation whereby a bit of excess is produced through the finessed interoperability of these dimensions and in relation to the surrounding environment. *Yoyū* is always a contextually embedded process. In the context of finessing the interval within the commuter train network, *yoyū* is produced within the gap between the principal *daiya* and the operational one.

For train drivers, producing *yoyū* demands a unique combination of intuition, technique, and attunement to the shifting conditions of operation. This can take years to master. For example, because of the lightweight composition of the commuter trains, load changes in relation to commuter congestion are particularly noticeable. Drivers must constantly adjust their braking techniques accordingly, since the heavier a train is, the more time and distance it requires to come to a full stop. If commuters were more aware of this, they might have a new level of respect for drivers, given the requirement for trains to stop precisely on a mark in each station. And, as Mito also suggests, they might have more tolerance for those instances of poorly executed braking in which the masses of densely packed commuters lurch forward simultaneously, producing a tremor that reverberates throughout the train.[36] Of course it is not only drivers who must master the combination of intuition, technique, and attunement to the shifting intensities of *yoyū*. Commuters embody *yoyū* as they inhabit the system's intervals as a tension between its emergent intensities and its sedimented patterns of relation (chapter 2).

Naturally, the practice of recovery driving is not specifically a postwar phenomenon. Drivers have employed it since the dawn of the railroad in Japan. What makes its application in the postwar period different is its tacit incorporation into regular operation. That is to say, *yoyū* became implicitly scripted into the planned traffic diagram as a means of overcoming material limits through a collective (human and machine) production of excess.

The Ticket Gate

Separating Tokyo from its commuter train network is unthinkable, yet neither is reducible to the other. The city and its complex train network must be thought of as two different systems of disparate orders of magnitude that have realized a certain functional coherence to operate together as a single associated milieu. That functional coherence is characterized not by seamless interaction but rather by a constant resolution of conflict between the different tempos and temporalities of the disparate orders. For this milieu to maintain its dynamic quality, the city and its commuter train network must remain distinct but entangled spaces.

Spatial delineation between the city and the commuter train network is maintained through the automated ticket gate. Just over one meter long, half a meter tall, and a hand-width wide, the machines stand in a row to form a porous border between commuter network and city. Until the early 1970s, in the Kansai region around Osaka, and the late 1980s, in the Kanto region around Tokyo, the ticket gate was

FIGURE 1.2. JNR Strike, Ikebukuro Station, 1978. Only one ticket puncher stands at the ticket gates.
Source: The *Yomiuri Shinbun*.

not automated. It was a small, oval enclosure manned by a uniformed railroad employee wielding a ticket puncher.[37]

The story of the automatic ticket gate's development is the subject of an episode in *Project X*, an immensely popular NHK (Japan's Public Broadcasting Corporation) documentary series.[38] Originally aired on NHK Television in June 2001, the series came in the wake of a decade scarred by persistent economic recession, unmistakable fissures in formerly idealized social institutions, and increasing juvenile violence. Against the backdrop of the sense of despair that hung over the preceding decade, the *Project X* series conveys a desire to reclaim the nation's present through the re-presentation of its accomplishments in the past.

Opening with footage from the 1960s of enormous crowds bottle-necked at train-station ticket gates manned by station employees, the *Project X* episode explains through narration that rapid economic growth in the era had created an enormous increase in the number of commuters in Tokyo and Osaka. As a result, ticket punchers at train-station gates were overwhelmed, which disrupted the system's flow while intensifying congestion and increasing the misery of the commute. This also led to a rise in the number of commuters falling from the platform onto the tracks. In order to reduce congestion to a manageable level, train stations were forced to periodically close the ticket gates, which had dire consequences with commuters, who were made to wait and thus possibly miss their trains.

In 1963, the program relates, the private Kansai railway company Kintetsu decided to solicit a technological solution for the problem from all the major Japanese electronic-appliance companies. The innovation, however, came not from one of the major companies but rather from a group of young and unknown engineers at Tateishi Denki, a small company on the verge of bankruptcy. Despite being brought to the edge of defeat by repeated failure and the withdrawal of financial support from Kintetsu, Tateishi Denki's engineers persevered and in 1967 produced the prototype of the contemporary automatic ticket gate.

The solution that Tateishi Denki eventually developed for the ticket exploited the same principle that a young technician at IBM was using at the time to create plastic identity cards for the CIA. All the necessary fare information for a monthly pass or a one-way ticket was encoded on magnetic tape attached to the backs of tickets and passes, which could then be read by an automatic reader in the ticket gate. After finding a company capable of affixing magnetic tape to paper, the problems remaining to be solved were minor, and the first automatic fare-collection system was born. In 1969, Kintetsu adopted the technology, and from the early 1970s onward the automatic ticket gate became a common fixture in subway and private railroad lines throughout Kansai. Only JNR continued to resist the technology nationwide (many critics claimed this resistance was due to union pressure to maintain a high number of employees). The new technology was finally adopted widely after the privatization of JNR by the new JR Group companies.

While facilitating operation beyond capacity, the automatic fare-collection system is important for the way in which it marked the first step in a process, one that has continued to evolve to this day, of re-casting the commuter as information, thus allowing for intensified logistical management of commuter bodies. Since the majority of commuters used a monthly pass, the new magnetic-tape commuter pass provided an unprecedented means for collecting daily ridership data, which could be applied to revising the *daiya* and to marketing advertisement space within train stations and train cars. Only a lack in computer processing power prevented wider application of the information and the culling of actual real-time train-usage data from the operation of the automatic ticket gates. In many ways, the introduction of the magnetic-tape commuter pass in the commuter train network is comparable to the advent of the bar code for commodity management in the early 1970s. As Jesse LeCavalier writes, the bar code revolutionized commodity storage and circulation at the time by allowing objects to be not only encoded with information but also treated as information within the logistical management of time and space.[39] The magnetic-tape commuter pass took that logic of information one step further through the link it provided to individual-commuter information collected at the time of purchase. However, the notion of exploiting that link for the calibration of advertising within the system would not arise until a number of economic and technological changes transformed the system several decades later (chapter 3).

Conclusion

Tokyo's packed commuter trains offer a spectacle that lends itself to a familiar narrative of the train as a medium of technology's dehumanizing, alienating, and rationalizing effects. In this chapter, by tracking the emergence of operation beyond capacity through the evolution of the tactic of finessing the interval, I have offered a different approach that emphasizes the packed commuter train as an expression of collective constituted around a margin of indeterminacy. My argument in this context has underscored how operation beyond capacity cannot be explained as simply the result of increased technological precision and efficiency. I show, rather, how it is the effect of a schema of operation that works to expand the network's margin of indeterminacy by decreasing the degree of scripted interaction between its parts and allowing instead for the finessing of intervals. In other words, somewhat counterintuitively, operation beyond capacity demands a mode of technological performance that is less rather than more determined. Accordingly, each expansion of the system's margin of indeterminacy for further operation beyond capacity produces a corollary increase in the dynamism of the network's collective environment. The underlying quality of these technological relations constitutes the technicity of Tokyo's commuter train network. The specificity of Tokyo's commuter train network is embodied in its technicity, particularly the precarious configuration of relationships articulated in the production of *yoyū*. My argument in this chapter has concentrated on the technical aspects in the emergence of a schema of finessing the interval while gesturing at the work of commuters. The following chapter will take up the latter in greater detail to explore how commuters finesse the interval to inhabit the margin of indeterminacy for operation beyond capacity.

Understanding operation beyond capacity in Tokyo's commuter train network through the evolution of its margin of indeterminacy challenges conventional models of techno-social analysis. The picture that this understanding offers of the train network as a dynamic technical ensemble whose functioning is contingent on its ability to enfold a great degree of indeterminacy goes against a dominant view of the railroad as an exemplary instantiation of a modern disciplinary apparatus. The terms *dynamic*, *open-ended*, and *emergent* are typically reserved for descriptions of the cybernetic-based information technologies of late-capitalist postindustrial society. By stressing the technicity of a technical ensemble—Tokyo's commuter train network—the argument I have provided here is meant to encourage an approach to questions about collective life that circumvents the assumptions at work in those descriptions.

NOTES

1. Adorno et al., *The Authoritarian Personality.*
2. Simondon, "On the Mode of Existence of Technical Objects."
3. Donald A. MacKenzie and Judy Wajcman take this latter approach of technology shaping society and society shaping technology in *The Social Shaping of Technology*, 2nd ed.
4. Félix Guattari, *Chaosmosis: An Ethico-aesthetic Paradigm.* Translated by Paul Bains and Julian Pefanis (Bloomington: Indiana University Press, 1995). See also LaMarre's explication of Guattari's concept in LaMarre, *The Anime Machine.*
5. Mackenzie, *Transductions*, 52–53.
6. As Adrian Mackenzie puts it, this approach asks us to understand a technological object lying "somewhere between a transient, unstable event and a durable, heavily reproduced structure"; see Mackenzie, *Transductions*, 14.
7. Simondon, like Alfred Whitehead, called this process of continuing individuation with the environment "concrescence."
8. Mackenzie, *Transductions.* See also Mackenzie's analysis of the margin of indeterminacy in ibid.
9. Simondon, *On the Mode of Existence of Technical Objects*, 17.
10. Deleuze and Guattari, *A Thousand Plateaus.*

11. One of the most extensive and detailed records is the eleven-volume collection produced by the Ministry of Transportation, *Nihon kokuyū tetsudō hyakunenshi* (ed. Nihon kokuyū tetsudô sōsaishitsu shūshika).

12. Nishiyama Takashi also attributes the economic boom during the war to the fact that the Japanese archipelago remained outside the battlefield, allowing industry to develop freely. Citing the definitive nineteen-volume collection on one hundred years of railway history, he points out that in 1914 JNR carried 166 million passengers and that by 1918 this number had risen to 245 million; see Takashi, "War, Peace, and Nonweapons Technology."

13. Fujii, "Intimate Alienation."

14. This accelerated the trend of the separation of domestic and work spaces that had been slowly emerging since the end of the Meiji Era (1868–1912). See Mito, *Teikoku hassha*.

15. Chalmers Johnson describes the adherence to a centralized model of planning at length in *Miti and the Japanese Miracle*.

16. Mito, *Teikoku hassha*, 123.

17. Taking the Chūō and Yamanote lines for example, between 1914 and 1919, ridership on the former rose from an average of 83,575 to 221,183 passengers per day and on the latter from 101,344 to 290,014 passengers per day. See *Shosen denshashi kōyō*, 227.

18. Mito, *Teikoku hassha*, 76–78. See also data on the number of train cars at the time for main lines in Tokyo in *Shosen denshashi kōyō*, 58–61.

19. Around this time companies began increasing the number of train cars per train from two up to four cars. See Denkisha kenkyukai, *Kokutetsu densha hattatsushi*.

20. Mito, *Teikoku hassha*, 76–78.

21. Ibid., 126.

22. See *Shosen denshashi kōyō*, 125.

23. Ibid.

24. Tsutsui, *Manufacturing Ideology*.

25. Denkisha, *Kokutetsu densha hattatsushi*. See also Mito, *Teikoku hassha*, 128.

26. Hein, "Growth Versus Success," 106.

27. Johnson, *Miti and the Japanese Miracle*, 200.

28. Yamamoto, *Technological Innovation*, 226. See also Johnson, *Miti and the Japanese Miracle*, 198.

29. Ivy, "Formations of Mass Culture."

30. Noda, Katsumasa, and Eichi, *Nihon no tetsudō*, 284.

31. Takashi, "War, Peace, and Nonweapons Technology." Distributing the propulsion system over the length of the train allows for increased acceleration and deceleration by eliminating the concentration of weight at one end of the train, in the locomotive, and by increasing traction. The system also decreases the maintenance costs and wear of the rail and roadbed because of the reduced weight; see Yamamoto, *Technological Innovation*.

Rationalization of the train system was also helped by Seiko corporation's introduction in 1952 of a special-edition clock with a second hand, which was quickly adopted for railway stations; see Mito, *Teikoku hassha*, 98.

32. Mito offers a compelling explanation for the close proximity of stations. She writes that Japan's modern cities inherited their station topography from the premodern city, which was a pedestrian society rather than a horse-and-buggy society due to restrictions that the Tokugawa shogunate put on the use of wheeled vehicles in order to curtail its vassals' military aspirations. Consequently, villages, inns, and resting facilities in the area, which later became train stations, were spaced fairly close together, as befits a pedestrian society. Mito, *Teikoku hassha*, 67–69.

33. But JNR's efforts to centralize really only took shape after the opening of the first Shinkansen (bullet train) in 1964. See Isamu, "Ressha shūchū seigyo sōchi (CTC) no kaihatsu."

34. *Nihon kokuyū tetsudō hyakunenshi*, 11.

35. Shinmura, Izuru. *Kōjien*, DVD-ROM (Tōkyō: Iwanami Shoten, 2008).

36. Mito, *Teikoku hassha*, 173.

37. The genesis of the automated ticket gate is a related thread in the technicity of Tokyo's commuter train network that I will elaborate on in a later chapter.

38. Akira Imai, director, *Prujekuto X chōsentachi: tsūkin rasshu o taiji se yo* [*Project X Challengers: Let's Eradicate the Commuter Rush*], DVD (Tokyo: NHK, 2001).

39. LeCavalier, *The Rule of Logistics*.

TWO

Inhabiting the Interval

When Henri Lefebvre identified the role of space in producing the social relations of capitalism, he transformed space from a passive background of historical materialism into a dialectically charged, active milieu rife with contradictions.[1] Providing a powerful intervention in Marxist theory, Lefebvre's work proved highly generative for thinking about such paradigmatic urban spaces as the commuter-train car as scenes of paradoxical mediation from which emerge, on the one hand, a modern, rational, urban subject compliant with the disciplinary imperatives of labor and, on the other hand, individuals highly attuned to the affective cues and desires of consumer economy.[2] Yet by rendering spaces of infrastructural mediation subject to the formal overarching logic of capital's and modernity's rationalizing processes, Lefebvre's approach runs into analytical limits. Left out of Lefebvre's conceptualization are the specific technicities of infrastructure and the particular quality of relations elicited within the human and machine collective.

How might thinking from within the margin of indeterminacy allow us to understand the commuter-train space, especially the practices and conditions of Tokyo's packed commuter train, differently? One point of entry into this question is through the phenomenon of the suddenly awakening commuter, which is when a commuter who appears to have fallen into a deep sleep suddenly awakens at a station and bolts from the train a split second before the doors close. This is something one typically encounters on a late-evening train. The commuter is in a state of absolute repose one moment—head thrown back, mouth open—and in the next instant frenetically animated, like a doll snapped to life with the yank of an invisible cord. Without so much as pausing to glance out the window and confirm the station, the suddenly awakening commuter has leapt to his or her feet, gathered a briefcase or bag, and rushed straight for the train door. As the train pulls away, one can see the commuter standing on the platform, squinting under the harsh fluorescent illumination, still half asleep and looking somewhat bewildered. Such moments break through the collective reserve, bringing faint smiles to the faces of those commuters remaining on the train. But such moments also instantiate a paradox that raises a question: How did the commuter know it was his or her station even while asleep? Or, to put it differently, how is it possible to be simultaneously asleep and seemingly awake and sensitive to the ambient fluctuations within the commuter train network?

Mito Yuko has an answer to this question. "The rhythm of the train is etched into the bodies of the city's inhabitants," Mito explained to me over tea during the first of several meetings in the posh lobby café of Shibuya's Cerulean Tower Hotel. To emphasize her point, Mito mimicked the train's cadence, slapping her hand on the table, "*kattak, kattak, kattak.* . . ." With frizzy hair, brown oversized plastic-rim glasses, and a slightly faded brown blazer, Mito's appearance was at odds with the crisp-suited salarymen and impeccably dressed middle-aged ladies sipping tea and coffee from the café's small name-brand designer cups at nearby tables. Her animated style of speech and fondness for sound effects were no less at variance with the subdued tone of conversation transpiring at those tables. Not surprisingly, our table drew glances. Mito ignored these as she went on to describe the process whereby the Tokyo commuter becomes deeply attuned to the rhythms of the system.

Essential to this process, she emphasized, is the "driving curve," which is a diagram mapping an ideal driving pattern of acceleration and deceleration between stations on a vertical axis of speed and driving time and a horizontal axis of distance. As a schema of the spatiotemporal division between stations and a supplement to the *daiya*, the driving curve enfolds a margin of indeterminacy that, in Mito's explanation, is the dimension through which one becomes a Tokyo commuter. Mito explained:

No matter what train line one takes in Tokyo, the pattern of acceleration and deceleration between stations is always similar. The pitch of the electric motor increases

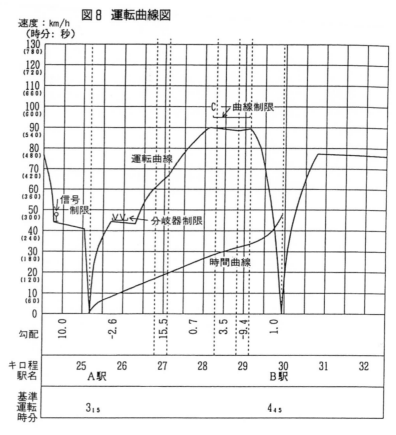

図 8 運転曲線図

速度：km/h
（時分：秒）

FIGURE 2.1. Driving Curve Diagram
Source: Okamura, Atsuhiro. "Ressha daiya no shōtai." *Tetsudō jyanyaru* 29, no. 6 (1995): 43–48.

as the train rapidly accelerates when it leaves the station: *eeeeewwwwwww.* . . . It levels off for a bit as the train reaches its cruising speed—*pweeeeeeehhhhh*—and then begins to drop as the train decelerates: *dreeweeeeeeeeh, tukatoo, tukatoo.* . . . [Then there's the announcement:] "We've arrived at such and such station." This pattern is the driving curve, and every commuter internalizes it from an early age. For commuters it's a soothing sensation, lulling them to sleep the moment they sit down on the train. Because the bodily rhythm of the city's inhabitants is in sync with this pattern, when there is a delay, even if it's only a matter of thirty seconds, they notice it. If the delay is more than a minute, they might actually begin to feel physiological discomfort [*shintai no seiriteki na fukaikan*].[3]

Mito's elaboration of the suddenly awakening commuter phenomenon exemplifies the kind of thinking with the margin of indeterminacy that is the aim of this chapter. In her description, the commuter is not formed by the train car's space but rather emerges within the tensions of the margin of indeterminacy that propagate in the rhythms, sounds, resonances, and kinetic energies of the system.

The commuter develops a deeply ingrained sensibility toward these modulating ambient intensities, which take on the quality of a living being who demands not mindless compliance but a corporeally bound labor of constant attention. Like a partner in an intimate relationship or complicated dance, the commuter becomes attuned to the system's tacit cues and sensitive to its fluctuations, so much so that the difference between thirty seconds and one minute registers viscerally as a change in mood that resonates as active discomfort. Accordingly, the commuter learns not only to inhabit the margin of indeterminacy but also to become comfortable within the precarious quality of the margin's tensions between structure and process, pattern and emergence. Thus the commuter is able to embody contradiction and finesse the interval of the commute so as to be asleep but remain awake, to be deeply attuned yet emotionally and socially detached. By finessing the interval, the commuter inhabits the margin of indeterminacy, creating *yoyū*—space where there is actually none—while exploiting the margin's patterns, energies, and variations.

Operation beyond capacity, this chapter argues, is not a system imposed on Tokyo's commuters; rather, it is a collective condition produced at an organic level in the labor of commuters to maintain the precarious metastable integrity of the margin of indeterminacy. Whereas the previous chapter focused on the genesis of the schema of finessing the interval for operation beyond capacity, this chapter explores how commuters finesse the interval to create a system that should not work but does—just barely—and in so doing come to inhabit a technosocial reality that should be both unbearable and unsustainable and yet is made to feel entirely normal. Confining itself to the space of the train car, this chapter considers how a collective emerges in this regard in conjunction with the labor of finessing the interval articulated in commuters' meticulous adherence to routine, the silence of the packed train, and the production of train manner. It considers as well the various forms of media—from cell phones to screens to posters—that have become essential features of the commute in recent years. Of central concern is how these devices become folded into the paradoxical mediations of the commuter-train car to inflect the limits and possibilities of its collective. What kinds of gaps, I ask, do they open within the margin of indeterminacy?[4]

In accordance with the overall aim of this book to think with the margin of indeterminacy of the commuter train network, the objective of this chapter is to think with the specific articulation of collective that is conjured in conjunction with finessing the interval within the spatiotemporal confines of the commuter-train car. This site is taken as a scene of mediation in which the qualities of collective life are made especially intelligible. In the space and time of the train car, we find the clearest expression of the tension between the forces of pattern and emergence, structure and contingency that are generated in the interaction between the system's internal technological organization and its associated milieu of commuters. This chapter thus brings the commuter forward not as a compliant body, but as an active body that has learned to think and act within the tension and precarity of the margin of indeterminacy of the commuter train.

The Packed Train

Japanese salarymen have a fixed schedule. I leave my house every day at exactly 7:05, arrive at the station and line up at the second door of the ninth car of the 7:23 commuter express. Every day exactly the same and always with a *Nikkei* [Economic] newspaper under my arm. From my station until Shibuya it's too packed to even lift my arms and hold the paper. But at Shibuya a lot of passengers get off and I have fifteen minutes to read all the important articles in the paper. There is a group of regulars I ride with but I've never spoken to them or exchanged a nod or greeting. A salaryman's energy is at the lowest in the morning. It's like "ach, back to work again and back on the packed train." You just have no energy to waste.
AKIRA, SALARIED BANK EMPLOYEE[5]

Akira's account distills a central paradox at work in finessing the interval, which is that while commuters grumble interminably about the unbearable and exhausting conditions of the packed commuter train, they pursue their daily commute on the packed train with meticulous if not zealous discipline as well as pride. Indeed, to be able to endure the conditions of the packed train day after day, sometimes for the entirety of a career, is to demonstrate one's strength as an adult and as a contributing member of society. Although collectively realized, such pride is individually claimed by recognizing without acknowledging the faces of commuters with whom one rides the packed train every morning. Like many commuters, Akira finds satisfaction as well in his capacity to optimize the brief fifteen-minute moment in his commuting time to prepare himself for the day by reading "all the important articles" in the newspaper.

It is worth noting that his depiction of congestion in his morning-train car via reference to being able to read the newspaper is not incidental. It is a standard metric that the Japan Association of Railroad Industries employs to illustrate the degree of immobility in relation to increasing levels of operation beyond capacity, as figure 2.2 shows. At 100 percent capacity, the illustration states, you can grab on to the hanging strap. At 150 percent capacity, you are shoulder to shoulder with other passengers but still have enough room to enjoy reading a newspaper. At 180 percent capacity, the newspaper has to be folded into fourths in order to be read. When capacity reaches 200 percent the pressure of bodies pressed together is particularly oppressive. An obstinate reader, however, might still find a way to read a weekly magazine. Anything above 200 percent, the illustration specifies, denotes an extreme condition and reading is impossible. As the train sways, the illustration explains, your body is forced to lean together with the others; you are entirely immobilized, unable to move even your hands.

Another level of paradox concerns the apparent, meticulous precision of Akira's schedule. That Akira is able to provide such a precise account of the motions and practices of his morning commute suggests conformity to a system of equal predictability and precision. Thus what Akira offers with his account is in many ways a very conventional image of the Tokyo commuter as an individual disciplined to comply with the operational imperatives of the technological system. But the inverse is actually more accurate, for when Akira leaves his home every day at exactly 7:05 a.m. to arrive at the station and line up at the second door of the ninth car of the 7:23 commuter express, the only

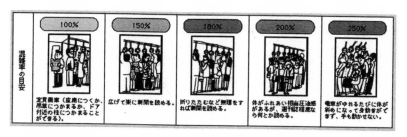

FIGURE 2.2. Congestion Illustration
Source: Mizoguchi, Masahito. "Nihon no tetsudō sharyō kōgyō ni tsuite." Japan Association of Rolling Stock Industries. 2007. http://www.tetsushako.or.jp/pdf/sharyo-kogyo.pdf.

certainty is that the 7:23 commuter express will not depart precisely at 7:23 a.m. Most likely it will arrive late and/or be delayed in departing. In other words, the discipline and fervor with which Akira pursues his daily-commute routine is an account not of the rigid exactitude of the system but rather of its margin of indeterminacy. It is the precarity of the system, not its certainty, that makes such zealousness and discipline of commuters like Akira necessary.

Silence

I knew all the faces of the people I commuted with. When they didn't show up a few mornings in a row, I'd start to get worried. And then when I'd seen them again I'd want to ask if they were okay and what happened. But I never did. If I had, I would have had to greet them every morning and probably would have eventually ended up sneaking off to ride in a different car so as to avoid them. It's just too much.

MICHIKO, FORMER PHARMACEUTICAL COMPANY EMPLOYEE[6]

Similar to Akira, Michiko recognizes the faces of her fellow commuters without ever committing to a gesture or action that could be taken as acknowledging their presence. What is more, Michiko's absolute aversion to social interaction with her fellow commuters despite the concern she confesses for them conveys a sense of ambivalence that has long been recognized as a fundamental quality of commuter space and the mark of a refined urban sociality among strangers.[7] The specificity of this ambivalence in Tokyo's commuter train network derives not only from the intense congestion within train cars that creates a proximity between commuters that, while appropriate for lovers, makes normal conversation impossible, but also, as both Michiko and Akira note, from the fact that one does not typically commute with strangers. The meticulous routine that commuters tend to follow, boarding the same train every day from the same train door, means that one's fellow commuters are not exactly strangers, yet also not acquaintances. The relational status is more indeterminate, suspended somewhere between familiar and unfamiliar, making them simply the "others" with whom one commutes.

Aversion to social interaction with those "others" stems, as Michiko suggests, from the desire not to expose oneself to the tedious and daily concern for reciprocal acknowledgement. At the same time, it also stems from a sense of consideration for one's fellow commuters—from a desire not to impose the tedious and daily concern for reciprocal acknowledgement on someone who would also most likely be forced to sneak off and find a different train car to ride. This sense of an under-lying connection with the others with whom one shares the intensely intimate space of the train car day after day, yet with whom one never converses, gives rise to a highly charged ambivalence that is rendered palpable in the absolute silence of the packed train. The silence of the packed train is not imposed on commuters; it is a condition born in the interplay between human and machine that constitutes part of the specific technicity of the commuter train network.

The silence of the packed commuter train derives in part from the muted quality of the apparatus. Moving from the train platform into the train car, one transitions from a dense soundscape of chimes, bells, reminders, warnings, and announcements into the hushed space of the train car. When the train doors close, the sounds of the plat-form become removed—not entirely gone, but muffled—as if they were part of a different world. Then the familiar revving of the motor takes over, and the station is gone. With the air-cushion suspension system that separates the train carriage from the chassis and with the tracks smoothed into single rails, an almost soundless gliding sensa-tion ensues. Gone are the jolts and shocks of the apparatus that once provided the ground for theoretical interventions into the habituated circumstances of technological modernity.[8] The sounds of the appara-tus are diminished to the soothing, modulating resonance of the ve-hicle's powerful electric motor and the soft whine of the oscillating air-conditioner fan. The silence amid such intense bodily proximity is such that one becomes aware of the sound of a fellow commuter's breathing, a stomach growling, and even sometimes a beating heart. Such bodily intensities propagate as a material force, producing a sensa-tion of contact that threatens the balance of commuters' assiduously maintained degree of (non)entanglement. It is punctuated only by the soft feminine voice of recorded service announcements and nasal station-call reminders from conductors.

As Michiko's and Akira's accounts suggest, the silence of Tokyo's packed commuter trains is a conscious and rigorously maintained pro-cess. It can be understood as a way in which commuters create space for themselves where no space actually exists. In the sense that it in-volves the creation of something from nothing, such space is analo-gous to the *yoyū* of finessing the interval (chapter 1). Where finessing the interval for operation beyond capacity creates a gap via an infor-mal dynamic of temporal debt and recovery, the silence of the packed train produces a gap through which commuters manage their degree of entanglement within the collective. While this is not an actual space, it is a palpable space nonetheless. The more intense the congestion of bodies within the packed train, the more intense the effort required to produce that gap. Consequently, and somewhat counterintuitively, the more one is immobilized by the pressure of surrounding bodies, the more one's body is actually in a state of what Thomas LaMarre calls a kind of "molecular" motion.[9] It is the pressure of such molecular motion that leaves commuters feeling exhausted.

Mediating Silence

The silence of the packed train is mediated by the dense advertising that inundates train cars in Tokyo, producing a visual register that commuters from other Japanese cities often find "too loud." Train-car advertising, or "train media" as it is called, is strategically positioned anywhere an eye might come to rest, if only for a second. Along with the usual line of posters on the curve between ceiling and wall that one finds in commuter trains throughout the world, at each of the four doors sets of hanging advertisements (*nakazuri kōkoku*) are suspended from the ceiling at the center of the train car. Occupying a premium position, these advertisements are typically monopolized by Japan's immensely powerful publishing industry and present textually dense displays of the tables of contents from various weekly and monthly magazines. They include everything from serious literary, economic, and social periodicals to mainstream tabloid entertainment and manga. Advertisements are also stuck on doors, wrapped around the standing pulls, attached to hanging straps, and sometimes even presented on the floor. In the last decade, advertising space was expanded to include the train exterior with so-called "wrapping advertisements."

Train media operates on a level beyond the conventional logic of distraction for a captive audience. While much of it is straightforward promotion of commodities and services, a lot of it is masterfully produced, blurring the line between advertisement and entertainment by exploiting clever phrases, puns, and captivating combinations of elegantly executed illustrations and photographs. Commuters tend to appreciate such work and even look forward to it, regardless of whether it compels them to purchase a specific commodity. Such anticipation and the satisfaction it provides endow the relation between the advertisement and the commuter with a sense of relationship. This is especially so with serialized advertisements.

Re-instantiating an age-old combination of mass transportation and mass media, serialized advertisements in trains are often stills from current television-advertisement campaigns telling stories that unfold, one installment at a time, around a product. One of the most successful recent examples is Softbank Mobile Carrier's "White" smartphone family-plan campaign, which features the ongoing tale of a quirky family of four. The mother, daughter, and son are played by known actors and media talent. The father, however, is a white Hokkaido dog. While conveying some message about the mobile carrier's pricing plan or technology, each commercial invariably focuses on the father, who thinks of himself as human and embodies the stereotypical Japanese father whose often conservative stance and stubbornness tend to get him into comically awkward social predicaments. The characters in such serialized advertising campaigns become familiar faces for commuters, while the ongoing progression of the serialized story takes on the feeling of a dialogue that stands in for the dialogue commuters cannot have among themselves. In the same way that viewers of a television series look forward to the next episode, commuters learn to look forward to character developments and resolutions of conflict in each installment. The extended and episodic nature of serialized advertisement also means that someone who returns to the trains after having been away for a time feels disagreeably out of sync with the environment. The sensation is not unlike trying to figure out the plots and characters of a film from the middle.

In recent years, JR East introduced flat digital screens above the doors inside the train. Between service announcements, these screens cycle through rebroadcasts of popular television commercials. When the JR East marketing group, JEKI, first proposed introducing the screens,

FIGURE 2.3. Softbank Mobile Carrier's "White" Smartphone Family Plan Advertisement. Courtesy of Softbank.

there was serious concern among train operators that the audio would disrupt the train's silence as well as interfere with announcements.[10] In a compromise, JEKI agreed to run the commercials without audio. Contrary to expectations, the soundless commercials proved remarkably effective in holding commuters' attention. Of course this makes sense if we consider the way dialogue between commuters and advertisements stands in for the dialogue that cannot take place among commuters. Having seen the commercials countless times on television, often in the context of a variety of morning-news shows that play an essential part of the commuter routine by providing a clock in the corner of the screen, commuters tend to be familiar with the commercial soundtrack and able to fill in the missing voices and melodies.

In accordance with the silence of the packed train, filling in the muted audio of the commercials is supposed to be a dialogue that transpires inside one's head. This is something that an individual I encountered on a mid-morning semi-packed Chūō Line to Shinjuku either did not understand or simply chose to ignore. Taking a position at the center of the car, directly in front of the screen so as to afford himself an optimal view of the looping commercials, the man treated the commuters in his vicinity to a perfectly executed reproduction of the audio from each commercial, modulating his voice as required to perform the various male and female characters. He was nothing short of an advertisement virtuoso, reproducing the nuance of every melody with perfect pitch. Aside from this odd behavior, there was nothing unusual about him. He was slightly heavy, wore a charcoal business suit over a white shirt and tie, and clutched a standard black nylon portfolio briefcase in his hand, giving him the look of an entirely ordinary young salaryman. But it was obviously not his first performance, and although he performed for himself, just loud enough for those nearest to him to hear, everyone who could hear moved away, leaving him at the center of a small gap that served to spotlight his behavior. Oblivious to or unconcerned by the attention, the man continued his performance without missing a beat, staring straight at the screen the whole time and falling quiet only in the moments between commercials taken up by service announcements.

Keitai: Connecting Silence

Nothing reshaped the silence of the packed train more than the introduction of the mobile phone, or keitai, in the 1990s. The keitai burst the silence of the packed train, transforming the space and time of transit from a scene of transitory disconnection into an ideal interval of communication. Although not developed exclusively for commuting, the *keitai* was designed specifically with the commuter crushed in the packed train in mind. Fashioned to be comfortably held and operated in one hand so as to leave the other hand free to hold a bar or hanging strap, the *keitai* provides each commuter with a discrete window not onto a passing landscape or city but into a dimension of communicative events and potential fantasy.

Keitai is short for *keitai denwa*, which literally means "portable phone." That people tend to abbreviate the term not only reflects a general penchant in Japan's culture for abbreviations; it also reflects, as the anthropologist Mizuko Ito wrote in 2005, the sociocultural specificity of technological imaginaries that distinguish *keitai* practices in Japan from mobile-phone use in other countries.[11] The *keitai*, Ito argues, is less about the notions of functional communication and mobility associated with the mobile phone in the United States and the United Kingdom, and more "about a snug and intimate technosocial tethering, a personal device supporting communications that are a constant, lightweight, and mundane presence in everyday life."[12] When Ito was writing, the *keitai* had been in popular use in Japan for already close to a decade, giving rise to a wide range of *keitai*-based practices, especially among youth, that researchers began to designate as part of a distinct *keitai* culture. Among the most notable *keitai* practices were *keitai* novels and *keitai* dating sites (*deai kei saito*).[13]

Most of these practices developed with the launch of the *I-mode keitai* in 1999, which gave users access to the mobile internet. *I-mode*, it is worth noting, does not provide the kind of full internet access offered by the smartphone, which became popular in Japan especially after Apple's release of the iPhone in the late 2000s. *I-mode* provides network access to a distinct internet domain, with *keitai*-specific sites and reduced data that allows for quick loading and minimal expense. This restriction was initially based in part on the limitations of wireless technology in the late 1990s and early 2000s. It was also not without its financial perks for *keitai* carriers. It allowed them to maintain tight control on content accessed via *keitai* while developing a slew of specific *keitai* mobile-internet services (like *I-mode* news or *Ezweb* train schedules and routing), for which carriers could charge users according to data packets sent and received. The *I-mode keitai* can be understood as a precursor to the smartphone as many of the practices spawned on the *keitai* have easily migrated to the smartphone.

While the kind of intimacy that Ito associates with the *keitai* has become generalizable for just about anywhere in the world with a robust wireless network and a thriving smartphone industry, its development began, to a large extent, with the advent of the *keitai* mail address (*keitai mēru*). This was a *keitai*-specific email address that one could only check via *keitai*, although the address could receive mail from a regular networked computer.[14] As such it offered a mode of discrete correspondence and connectivity through a single device that one always kept close to one's body, either in one hand during the commute or in a pocket so as to be able to feel the vibration announcing the arrival of a new message. Users could use the *keitai* mail address to subscribe to *keitai*-specific content (such as mail magazines, *keitai* novels, *keitai* news, and message boards), which would then be delivered daily to one's inbox. In contrast to content accessed on the internet, via a smartphone, or via a computer, content that arrived in a mail inbox took on a sense of personal address, which lent the *keitai* an affective aura of intimacy.

Nowhere is the intimacy of the *keitai* more pronounced, more amplified, and more important than in the packed train. Text, not telephony, is the preferred mode of communication, as speaking on the phone in public, especially in the train, is generally shunned in urban Japan. Within the silence of the packed train, communicating by text becomes like whispering to oneself, spurring one to share personal feelings.

Train Manner

The silence of the packed train, as an expression of the *yoyū* that commuters summon to modulate their degrees of entanglement with one another, reflects what Marc Augé, in his ethnography of the Paris Métro, calls a collection of "solitudes."[15] At the same time, the silence reflects commuters' recognition of not only the collective nature of this solitude but also the precarious nature of its process of finessing the interval. In yet another articulation of the paradoxical mediations at work in the packed train, silence is an underlying mode of embodied communication that makes operation beyond capacity possible. As such, it is a paramount expression of what is called in Japan "train manner" (*densha no manā*).

In the simplest sense, train manner refers to the behavioral conventions by which individual commuters negotiate the network as an embodied public space and technological environment. When commuters choose to text rather than talk on their smartphones, or to yield seats to pregnant or elderly passengers, although those actions clearly reflect general social values, they are also inseparable from the maintenance of the integrity of the margin of indeterminacy. Thus train manner denotes the indivisibility of the technological and social conditions constituting the commuter train network. It is also categorically distinct from law. Laws in the commuter train network cover three carefully stipulated categories of violent behavior: groping (*chikan*), acts of destruction/sabotage (*hakai kōi*), and physical violence (*bōryoku*). The exercise of authority in overt instances of violence tends to be swift and precise, with railroad employees mobilized in overwhelming numbers to intercept and remove offenders from a train. Violations of manner, by contrast, invoke no legal repercussion or punishment. They also rarely prompt confrontation, since commuters tend instead to accommodate disruption through passive tactics such as averting attention or even moving away.

Train manner is a performative process in which commuters embody, as individual commuters and as a collective, the commuter train network constituted in finessing the interval for operation beyond capacity. The specificity of train manner in comparison to expressions of rider etiquette that one finds in transportation systems throughout the world, and even in other Japanese cities, derives from the highly precarious quality of the Tokyo commuter train network's technicity. This precariousness tends to make train manner an ongoing subject of informal debate among Tokyoites, whether in casual conversation over drinks or as lengthy, heated discussion threads on websites and social-media communities. Typically up for discussion in these forums is the question of what exactly constitutes train manner. For some, train manner is a matter of basic common sense. For others, it is an expression of compassion or a manifestation of elementary empathy evidenced by an individual's capacity to read social cues and respond accordingly. No matter the opinion, everyone has a train-manner story. Sometimes these are humorous accounts of encounters with eccentric individuals or behavior on the train. More often, train-manner stories are indignant accounts of perceived infractions of train manner. These tend to be somewhat predictable stories about commuters cutting the platform queue, young women applying makeup on the train, or individuals ignoring priority-seating norms. Conveyed with urgency, disbelief, and exasperation, stories of train-manner infraction become historical allegory for claims of the rapid deterioration of the underlying social fabric. Train manner, according to these stories, is always in a perpetual state of crisis.

Such anecdotal accounts of manner call attention to the commuter collective as a scene constituted in indeterminacy. But they do not provide an avenue into the workings of manner, as it were, nor a clue as to what other work manner might be doing. To understand these, we have to look more closely at the relation of manner to the performative processes of the collective constituted within the margin of indeterminacy under operation beyond capacity.

The Train-Manner Poster

The relation between the margin of indeterminacy and the commuter collective is rendered visible in the train-manner poster. Train-manner posters are ubiquitous within Tokyo's commuter train network, appearing in stations, on platforms, and in trains. Some posters deal with the system's mundane operational requirements, asking commuters to queue and board in an orderly fashion, abide by priority-seating signs, not lie down on the seats or floor, and so on. Other posters attend to the quality of commuter experience, requesting that commuters not use cell phones, that they lower the volume on personal audio devices so that music does not "leak" (*moreru*) from the earbuds, or that women not apply makeup while on the train. Such posters are quasi-institutional iterations: although their production and distribution is sponsored by the train company, the matter of which specific manner issue the posters should address is often solicited from among commuters or youth groups, sometimes as part of a competition for the best manner poster.

Importantly, train-manner posters do not produce train manner. They are not disciplinary technologies that effect commuter compliance to behavioral standards. According to Yoshihara Mihoko, the Senior Project Director of Planning and Production at JR East's train media and advertising group (JEKI) in 2005, research shows no evidence of a direct corollary between conformity to manner and the manner poster.[16] Commuters do not simply fall into line, as it were, the moment the posters go up. Rather, the manner poster *elicits* a sense of a collective by calling attention to the existence of manner through its violations. In so doing, manner posters invoke the collective as a precarious assemblage of human and machine formed in the paradoxical mediations of the system's intervals.

Strategies of elicitation have shifted over the decades, but this basic premise has not. Consider, for example, a train-manner poster that appeared in train stations throughout Tokyo in 1925. Produced by Japan's railroad authority under the header "Alighting Passengers Have Priority. Please Board Orderly," the poster carries the following explanation:

A delayed departure of 30 seconds at one station amounts to a delay of 7 minutes across the 14 stations between Nakano and Tokyo, and a delay of 3 minutes and 30 seconds across the 7 stations between Kamata and Tokyo, resulting in a reduction in the number of trains during the 2 hours of evening and morning congestion between Nakano and Tokyo from 44 to 35 and between Kamata and Tokyo from 20 to 17, which increases congestion on trains by an extra 2000 to 3000 people.[17]

One wonders how many commuters might have paused before the poster, perhaps even producing a pencil and paper from a briefcase or pocket to assist in calculating the logic behind this delay scenario. Despite the injunctive tone adopted in the heading, the poster is not an expression of "training" under a disciplinary regime that decrees, "Obey, or else!" Rather, it is more an entreaty for compliance that elicits commuter cooperation through a rational demonstration of the significance of the gap between the planned *daiya* and the operational one. Notably, after the first line there is no explicit addressee here, just a statement of operational logic. That logic conveys the operation of the gap as a material force by showing how a mere thirty-second delay metastasizes in the spatiotemporal intervals of stations and trains into two- to three-thousand extra bodies. What begins, then, as a straightforward and pragmatic display of system logic quickly degenerates into a hysteric crowd scenario as the cumulative chain reaction of disruptions, laid out as consecutive clauses separated only by commas, leaves the reader breathless and confused.

Contemporary manner posters tend to take a different approach in eliciting collectivity, although the underlying logic is the same. Whereas the poster from 1925 foregrounds intervals between stations and trains, contemporary posters translate intervals into a tension between affective and prescriptive registers of the commuting experience. Consider, for example, a manner poster from JR East's Chūō Line with the heading "Line Up Precisely When Boarding the Train, *ne*" (*jōsha no sai wa, kichin to narande, ne*), shown in figure 2.4. The next line is more specific: "In Order to Facilitate Smooth Boarding, Please Cooperate by Forming 'Orderly Lines'" (*sumūzu ni noreru "seiretsu jōsha" ni go kyōryoku kudasai*).

FIGURE 2.4. Line Up Here Manner Poster. Courtesy of JR East.

Below this entreaty is an illustration that puts the viewer at the platform level, looking at the legs of two commuters. The legs are perfectly centered within a white bracket marking one of the designated queuing points behind the yellow line on the platform. One of the commuters is dressed in slacks and cumbersome brown shoes—presumably a man. The other, presumably a woman, is wearing a skirt, stockings, and pink polka-dotted shoes. Each of the inside shoes is animated with a conventional, gender-normative smiley face. The man's shoe has a big, happy, "masculine" smile, and the woman's shoe a more demure, half-circle, "feminine" smile. The woman's shoe is also winking. Behind them, on the opposite side of the platform, we see a train with its doors closed. The illustration of the apparatus is absolutely accurate, yet softened with green and greenish-blue hues such that it is not photorealistic.

Gone is the rational exposition of system logic that we see in the poster from 1925. Instead, the poster solicits cooperation in managing the system's intervals through a combination of affective and prescriptive registers, tacking back and forth between directive statements, mimetic gestures, tonal inflections, and aesthetics of cuteness. Its target of address is simultaneously personal and impersonal: "you" the individual and "we" the general commuter body. The poster's prescriptive solicitation begins with the addition of *"ne"* at the end of the first line. Notoriously difficult to translate, *"ne"* commonly works to mediate cooperation or confirmation in social interaction. Following the imperative form, it suggests a "positive politeness strategy" that is further moderated by its association with female speech.[18] Its appearance at the end of the first line works to transform the statement from an imperative into an indeterminate petition that is softened by the implicit evocation of a female tone and the slightly coquettish wink from the "female" shoe. The use of the English term "smooth," written in Katakana as (*sumūzu*), instead of the more bureaucratically inflected Japanese equivalent *junchō* in Kanji, injects a tone of playfulness that again works to soften the ostensible directive. In a slightly more passive-aggressive manner, enclosure of *seiretsu jōsha* (lining up in an orderly fashion to board) in quotation marks designates the stipulated action as an established practice whose precedent *should be* part of common knowledge. Modeling the instruction, the illustration communicates through embodied gestures that encourage mimesis. At the same time, the smiley face and winking shoes exploit tropes of playfulness and cuteness that supplement the poster's overall strategy of firm persuasion. Finally, all pretense of injunctive authority is deferred in the fine print in the bottom section explaining that the poster was produced by the "Railroad Youth Group" (*tetsudō shōnen dan*) by their own accord and out of concern for the wellness of society. Although not entirely inaccurate, it leaves out the fact that manner posters such as this are often the result of competitions or direct solicitations to commuters for manner-theme drawings.

Using a combination of affective and prescriptive registers to make such an appeal is standard for train-manner posters. Even posters that take a more aggressive stance with admonishing phrases like "Don't pretend to be sleeping just to avoid giving up your seat to an elderly person" tack back and forth between directive, gesture, and illustration, which work simultaneously to evoke the individual and the collective as addressees. This is especially clear in manner posters that employ more playful forms of address. JR East's vegetable-pun manner-poster series provides a good example. "You Know it's Faster When You Line Up!" (*naranda hōga haya[imo]n*) suggests a poster in the series, depicting cute, smiling potatoes queued in an orderly fashion on the platform (figure 2.5). The pun derives from the phonetic commensurability between the word for potato (*imo*) and the suffix *mon* added to fast (*hayai*).

Through its combination of attenuated directives, illustrated gestures, and modulated tones, the train-manner poster instantiates what William Mazzarella calls "affective management."[19] Denoting the "professional coordination of affect," the term initially calls attention to the necessity for public discourse to rely on more than just disciplinary power to produce collective life. As Mazzarella writes, "Any social project that is not imposed through force alone must be affective in order to be effective."[20] Mazzarella's intention is not to collapse affect

FIGURE 2.5. Manner Potatoes. Courtesy of JR East.

into discourse. Rather, in a move that has significant ramifications for thinking about train manner and the space of the train, he argues that collective life emerges in the precarious modulation within the gap between affect and symbolic elaboration, or "affect and articulation." Collective life, that is, is an "incomplete, unstable, and provisional" process, a forever-unfinished dialectic tacking back and forth between the symbolic and affective mediations of public discourse. As Mazzarella puts it, the former "is abstract and pertains to the formal, legal assemblage of citizenship and civil society. The other [affective mediation] gets us in the gut: it is equally impersonal but also shockingly intimate, and solicits us as embodied members of a sensuous social order."[21]

In calling attention to the gap as a scene of processual precarity in which collective life emerges, Mazzarella essentially invites us to think about power as topologically constituted, immanent in the lacunas of mediation and eventuating in the folding of its incommensurable registers. The gap, he writes, becomes the "condition of power's efficacy, if by efficacy we mean its capacity to harness our attention, our engagement, and our desire."[22] In other words, what compels our return to and investment in the processes of collective making is not merely the subjugating force of institutional discourse or economic or technological forces. Participation in the collective is elicited from within the gap itself as a space of process and potential, a scene in which one perceives the chance to try again and again at becoming someone or something different.

For Mazzarella, the gap eventuates in the visual and audio mediations of public discourse, including technologies of the state, of the entertainment media, and of consumer advertising. In the commuter train network, the gap forms in the lived experience of the network's margin of indeterminacy and the tension the margin produces between embodied intensities and prescriptive iterations. The embodiment emerges with the modulating resonance of the train's electric motors and in the peculiarly intimate yet detached sociality of the silent, packed train. The intensities of commuting arise within bodies compressed together so tightly that commuters can hardly breathe; within the fluorescent glare; and within ceiling fans that stir up a thick air imbued with the flowery fragrances of shampoo, perfumes, and body soaps along with the miasmic vapors of stale cigarette smoke, coffee, sweat, and bodies in various stages of stress and decay. Such embodied, nonrepresentational intensities of the network are in perpetual tension with the sedimented practices of commuting that inform the tacit rules of interaction and the announcements, reminders, and warnings that are constantly broadcast throughout the system.

The commuter collective emerges as the performative process whereby commuters as individuals and as a group manage the gap between the affective and prescriptive registers of the commuting experience. Train manner, it is important to reiterate, is an expression of this process, not the producer of it. It is the expression of an operation of power that is topologically constituted, which is to say that participation in the packed train is elicited from within the gap itself as a scene of possibility and potential. Thus when commuters declare "there is no choice" but to ride the packed train day after day (chapter 1), their statement is only partially accurate. The structural circumstances of life in Japan's advanced, capitalist society do indeed compel individuals to take their place every morning on the packed train in order to arrive in a timely fashion at their places of employment and educational institutions. At the same time, what also brings commuters back to the packed train day after day is the immanent sense of promise produced in the highly precarious process of finessing the interval for operation beyond capacity. The promise of the margin of indeterminacy is instantiated in its always tentative realization of a collective capable of overcoming its material limits. To ride the packed train is to be part of a collective, part of a process that no other commuter population in the world seems able to emulate or endure. It is this promise of collectivity that turns incidents and stories of infringements of train manner into such an urgent topic of discussion as evidence of the impending collapse of social life in Japan.

The Erotic Economy of the Train Car

If the precarious process of finessing the interval for operation beyond capacity turns the packed train into a space that harbors a sense of promise, the nature of that promise is not the same for female commuters and male commuters. For female commuters, the packed train cannot be disassociated from the always present threat of sexual violence that is enabled in part by the historical gendering of the train car as a space of male labor and fantasy. While female commuters have always existed, their stance as commuters has reflected their place within the market as precarious laborers whose contributions to home and nation have been essential for the financial integrity of both yet seldom recognized as important. As considerable literature on gender and labor in postwar Japan have shown, women constituted a crucial part of the labor force behind Japan's so-called "economic miracle," but they tended to be confined to nonmanagerial positions and defined as part-time labor within the tier of small- to medium-sized companies and businesses supplying parts or specialized services within large conglomerates, at the center of which was always a major bank and large company.[23] This made women a cheaper source of labor, as companies could forgo providing their workers with benefits, opportunities for advancement, and wage increases. This also made it possible for companies to quickly add or shed workers in accord with the waxing and waning of production demand from the first tier of large companies. This system was aided by the middle-class ideal that a woman should quit her job after marriage and devote herself to domestic life.

While this combination of circumstances worked to render women a structurally invisible sector of the working population, it also promoted an association of the packed commuter train with the salaryman life and the organization of the train-car space around normative notions of heterosexual male desire. One need only look at literature from the early twentieth century to realize that the construction of the commuter-train space as a scene of male labor and desire dates back to the emergence of commuter culture in Japan.[24] If such literature is evidence that women have always been part of the commuter population, it is also allegorical of the structural inequalities of the labor system, which requires women as both objects of desire and laboring bodies, yet works to obfuscate their role in regards to the latter.

Despite considerable train-car advertising directed at female commuters, the organization of the train-car space as a scene of male heterosexual fantasy persists, especially through the visual economy of advertisements. While advertisements directed at female commuters attend to things like fashion, hygiene, cosmetics, health, and even domestic life, there is little that might be called erotic. By contrast, at any given time, one finds countless advertisements for weekly magazines and manga that invariably offset their dense textual registers announcing the latest government scandal, social crisis, or crime with a glossy color image of a woman in a bikini (a *gurabia idoru* image). The woman is usually in her late teens, rarely past her early 20s, and typically shot in a pose that emphasizes a buxom figure.

Such advertisements are at the core of the construction of the train-car space as a scene of heterosexual male fantasy. They are always large and hang in areas occupying prime visual real estate in the center of the train car over commuters' heads. In terms of their relation to the magazine's table of contents, the images suggest that male participation in concerns of civil society, no matter how sensationally pitched, requires the additional promise of the erotic perk that comes with the magazine's section of full-body, glossy images of young women in sexually suggestive poses. Amid the close company of the crowded train, the male reader's glance at these first pages (I have never seen a woman looking at the section) is always furtive, just slightly slower than the usual time required to turn the page.

A more extreme example of the erotic visual economy centered on heterosexual males that permeates train space is so-called "adult manga" (*adaruto manga*), which one finds on magazine racks beside the conventional fashion, cooking, and news magazines at convenience stores in and around train stations. Although the advertising industry is prohibited by censorship rules from adorning magazine covers with anything more explicit than a swimsuit photo, adult-manga magazines have recourse to illustrations, which they make use of in sometimes highly creative ways. Along with images of the bikini-clad woman and the indelicate illustrations in adult manga, the male-oriented erotic visual economy includes explicit pictures of young women in the evening sports paper. No comparable images or illustrations of men are produced for the female commuter population.

The Train-Car Pervert

Is there a relation between this erotic visual economy and the chronic violence in packed trains against female commuters perpetuated by molesters, or perverts (*chikan*), as they are called? Although I have found no research exploring this question, it is hard to imagine that no corollary relationship exists. Much has been written about the problem of sexual violence against women in trains from a sociological perspective.[25] This work has been helpful in making the problem visible. Attending to the issue of the molestation of women on trains would require a book of its own, and it is not my intention to address the problem in depth here. Nevertheless, the erotic images in trains raise questions that must be addressed.

That the construction of the commuter-train space as a site of heterosexual male desire slides problematically toward violence against women is supported by the fact that the sexual-entertainment industry (*fūzoku*) in Japan, which operates entirely as a legal "health" service, explicitly capitalizes on the connection with simulated commuter-train scenes for men to pursue groping fantasies.[26] Such simulations of "pervert commuter train" (*chikan densha*) cater specifically to men, providing the fantasy, as one website for the service describes it, of "forcefully groping defenseless women" (*mubōbi onna o kyōsei chikan*). There are no comparable services that I know of offered to women.

The molestation of female commuters has been a problem on trains in Japan since the emergence of crowded trains around the turn of the twentieth century. Critical representations of the problem, whether in Japanese cinema, literature, or television dramas, tend to present it as an issue of schoolgirls being molested by older male commuters, but the problem plagues commuting women throughout their careers. As early as 1912, some train companies even initiated "women-only train cars" in an attempt to mitigate the violence.[27] In contemporary Tokyo, this tactic continues with a certain number of designated "women-only train cars" on trains during rush hours.[28] Train companies have also long deployed posters in railway stations and trains to remind commuters that groping is illegal. The ones I saw reminded commuters that "molesting, destruction of property, and violence are criminal acts punishable by law" while also instructing women to grab the hand of their assailant, raise it in the air and yell "*chikan*" (pervert). The train company's decision to include "destruction of property" together with molesting and violence in the same poster gives one pause. It reflects more than just a lack of tactfulness on the side of the train company, pointing instead to the possibility that the persistent problem of the molestation of female commuters on trains stems from an inability or even unwillingness within a male-dominated corporate world to reflect on and tackle deeper sociological presuppositions enabling violence against women.

According to Yoshihara at JEKI, about once a month JR East receives a complaint concerning *gurabia idoru* images in its trains. Usually, Yoshihara explained, the complaint is from a mother concerned about her daughter being exposed to such demeaning representations of women while commuting to school. But because there are so few complaints, JR East has never felt compelled to take action. Although Yoshihara hesitated to link such images with the violence of groping, she recalled that in the past feminist groups had opposed the bikini pictures in trains for precisely that reason—also to no avail, as their claims were never recognized.

The connection between the erotic visual economy of the train car and the molestation of female passengers seems to go unrecognized for the same reason that JR East does not feel compelled to respond to the few complaints that it receives: namely, many incidents of groping go unreported. Female commuters with whom I spoke, including both acquaintances and friends, sometimes explained that with the intense compression of bodies in the packed train, they often felt uncertain whether the hand that brushed or grabbed them did so with intention.[29] But it was not the uncertainty that silenced them; rather, it was the silence. As Yuko, a twenty-seven-year-old female administrator at an importing firm who commutes almost an hour each way on the Chūō Line, confided, while there were times when she suspected someone of touching her, she did not feel confident enough to carry out the required action of grabbing the groper's hand, raising it up, and yelling "pervert." She was afraid of the effect such an action would have in the silence of the packed train. For Yuko, riding in the women-only train car is not an option because of where it would place her on the platform at her destination and the crowds she would have to navigate to arrive at work on time.[30]

Finessing the Final Interval

The evening rush hours can be divided into distinct groups, suggests Mito Yuko.[31] There is the "punctually return home" group between 6:00 p.m. and 8:00 p.m., the "drink before going home" group from around 8:00 p.m. to around 9:30 p.m., and the "return just before the last train" group after 11:00 p.m. One outstanding feature of trains in Tokyo is that from around 9:00 p.m. until the last train, which is just after midnight, train cars on main lines tend to be highly congested, although usually not to the level of the morning rush. But the evening commute home offers an experience almost antithetical to the morning commute.

The smell of alcohol hangs in the air inside the train at night. Men in suits, their faces flushed from drinking at obligatory business-entertainment outings or over snacks with colleagues or friends around the station, hang by one hand from the standing straps in an effort to remain upright on rubbery legs. If the colleague or friend happens to live on the same train line, the conversation, relieved of any inhibitions by the alcohol, continues in the train in an animated spirit until one of the men reaches his stop. Similarly, men and women in their twenties and thirties and groups of university-age students in modish but casual dress try to maintain the structures of their small groups despite the dispersive pressure of the crowd. They rehash the night's events or share stories about incidents at work, laughing loudly at times. Those who are alone, which is often the majority, hang quietly by the standing strap or, if lucky, savor the respite offered by a space on the soft, cushioned benches lining the sides of the train. Some thumb email messages or scroll through pages of minute text on *keitai*. Others are absorbed in pocket-sized paperback novels or flip-scan slowly through thick magazines or manga. All emit an aura of exhaustion. The long hours under the fluorescent glow of the office lights and the effort required to be deferential toward a colleague or superior over the course of the evening at a noisy, smoke-filled restaurant-bar are easily discernible on their faces.

Packed among the warm bodies and comforted by the soothing rhythm of the train, people find their eyelids growing heavy. But sleep is only really possible for those lucky enough to have procured a seat on the train. Exhausted, they surrender to the comfort of the modulating resonance produced by the train's powerful motor. Books or cell phones still in their hands now fall to their laps, open to a page half-read or an email unfinished, as their heads begin to nod. The question of what to do when the person sitting next to you on the train begins to doze off, and with each sleepy nod a head comes closer to resting snugly on your shoulder and the full weight of an upper torso increasingly leans against you, is a common dilemma, born of the ambivalent character of the train-car space and time, for the night commuter.

It is also not unusual in the mid- to late evening to come across a train car that has been mostly abandoned by passengers because a drunken passenger has vomited in it. Passengers are not the only ones who need to worry about this: platform attendants need to be especially on guard. Emboldened by alcohol, commuters show no fear in using the narrow corridor on the wrong side of the yellow line at the platform edge to bypass the crowds queued for the train. Tensions rise as the train approaches and platform attendants—wireless microphones in hand—rush to urge passengers to clear the danger zone. Their voices, amplified over the platform speakers but competing with the countless sounds of the system, convey increasing apprehension as the train enters the station. The scene is endlessly repeated, so much so that train companies have taken to trying to reach commuters pre-inebriation with posters imploring, "Let's return home safely on days that we drink."

For train drivers, the inebriated commuter is forever a source of anxiety. "Evenings are the hardest shift. All I ever think about is getting through the hours without an accident," confided Saito, a veteran driver of twenty-five years on JR East's Yamanote Line who seemed nervous and just generally uncomfortable during the hour we spoke for at a Doutor café near Shinjuku Station.[32] In those twenty-five years, Saito had only one instance of a commuter falling from the platform to the tracks, but the individual was particularly lucky and managed to escape to the side just in time. In 2001, two men, one of whom was an exchange student from South Korea, jumped to the tracks on the Yamanote Line at Shin-Okubo Station (just outside Shinjuku) to save a drunken fifty-year-old salaryman. The rescue attempt was unsuccessful, and all three men were killed by the arriving train. Although platforms are equipped with clearly labeled and strategically placed emergency buttons to stop trains in such circumstances, either no one thought to push one or everyone was too engrossed in the drama occurring on the tracks to remember.

The End of the Line

For every commuter who awakes miraculously in time to leap from the train at the right station, there is another who does not awake in time. Some miss their station by a few stops. Others awake only at the end of the line. For members of the "return just before the last train" after 11:00 p.m. group who find themselves stranded at a station in the city, there is nothing to do but fill the gap in time before train service resumes early the next morning. They become transit refugees, in search of yet another space of in-between-ness to endure the long transition to morning. One finds these commuter refugees hunched over tables, nursing cups of coffee, in the twenty-four-hour fast-food chains and family restaurants invariably located in convenient proximity to the train station. They are a solitary and tired-looking lot, often struggling to shake off the alcohol consumed earlier with colleagues and convey a convincing performance of wakefulness so as to not be asked to leave their twenty-four-hour refuge. Those transit refugees with a bit of cash to spare might shelter at an all-night manga café or in a private booth at an internet café. While the former promises an abundance of manga neatly ordered on long shelves, the latter promises connectivity from the comfort of reclining and aggressively motorized massage chairs.

Among the many places within the network to find oneself stranded until morning, the most appealing is perhaps Takao Station, at the foothills of Takao Mountain on the far western end of the Chūō Line. Ten minutes by foot from the station, an all-night spa caters at night to the weary transit refugee. The building is a large, blunt, and somewhat worn-down concrete structure painted inside and out in a tone-deaf color scheme that lends the whole place the depressing and ponderous ambience of a Soviet-era resort. The spa's icon is a smiling egg-shaped person with stubby arms and legs; three squiggly lines rising from its head are supposed to represent steam but look more like thinning hair. Displayed on the facility's front windows and doors, this icon does little to lift the somewhat gloomy atmosphere of the building.

Nevertheless, the baths and other amenities offer just what any tired transit refugee would need to recuperate and return to the packed train the next morning. For a mere 1,080 yen (around 10 US dollars) to enter and another 1,080 yen to stay until morning, one receives a robe to change into and access to both a sauna and a range of hot and medicinal baths. One of the baths is outside in a small garden, while the others are in a cavernous bath hall. A small restaurant within the facility stays open until 1:30 a.m., serving a basic fare of noodle and rice dishes, beer, and snacks, such as dried squid and crackers, to go with the beer. After the restaurant closes, customers can still purchase food and drink from a line of machines, which offer everything from canned pop and coffee to large cans of beer, as well as snacks. Outside the restaurant, a line of recliner chairs faces a giant window with a door that leads out to a balcony. In another room, rows of soft reclining massage chairs all line up facing the same direction. Each chair has a small television stand in front of it, which makes the whole setting appear like the inside of some aerodynamically impossible first-class section of an airplane, or maybe a media room on a drab second-rate cruise ship. For those too agitated to sleep, there's a game room with the typical array of arcade games that cycle through noisy demonstration scenarios all night regardless of whether anyone is in the room.

Although the majority of the customers in the facility tend to be middle-aged men, there are always a few female customers as well. There is little discourse or mutual acknowledgment of each other's existence. If customers are not sleeping, they are absorbed in manga or newspapers or even sitting silently over a can of beer. Conversation is not only inappropriate but entirely unnecessary. People keep to themselves, sharing the intimacy of the bath and various recuperative apparatuses in a silence that feels uncanny for the way in which it recalls the packed train.

Conclusion

Thinking from within the gap materialized in a train car in Tokyo's commuter train network allows us to view the particular quality of the relations of the network's margin of indeterminacy. In so doing, we see that operation beyond capacity is a collective labor of human and machine. For commuters, that labor involves negotiating the tensions that emerge in the repetitions and differences that define the daily commute. The former is articulated in the sedimented patterns of interaction and movement within the network—the familiar faces with whom one lines up each day at the same time and at the same train door, the silence of the packed train, the cycle of announcements and advertisements, and the comforting rev and pitch of the powerful electric motors as the train accelerates and decelerates between stations. The latter is produced in the delays, irregularities, and fluctuations that the network must enfold in its milieu in order to accommodate commuters far beyond its infrastructural capacity. To become a commuter is to cultivate an active, not compliant, body attentive to the tensions and paradoxes of the margin of indeterminacy and invested in the reproduction of its collective order. At the same time, becoming a commuter involves learning to finesse the patterns, energies, and variations within the margin of indeterminacy so as to create space for oneself without compromising the overall collective order.

The commuter's capacity to inhabit the margin of indeterminacy reflects an ability to affect and be affected by machines, an ability that carries both positive and negative values. On the one hand, it demonstrates a potential for ontological entanglement with technological process, which is the foundational condition of possibility for collective life. Human beings do not have a monopoly on the potential for ontological entanglement with technological processes. As anthropologists have shown, there are plenty of nonhuman organisms that have learned to inhabit technological infrastructure quite well.[33] What human beings possess, however, is a far more expansive capacity for becoming with technology that goes far beyond merely adapting and involves instead extensive conceptual and material innovation. On the other hand, it is precisely that capacity for such expansive adaptation that becomes problematic in that it allows us to not only accommodate technology but also learn to inhabit quite comfortably forms of collective life that are not conducive for our own flourishing nor the flourishing of other life on this planet.

Notes

1. Lefebvre, *The Production of Space.*
2. For a wonderfully insightful historical analysis of the commuter-train car in Japan in the early twentieth century, see Fujii, "Intimate Alienation."
3. Mito Yuko, interview by the author, July 2, 2005, in Shinbuya at the Cerulean Tower Hotel.
4. This question is informed by Thomas LaMarre's observation, cited in the introduction, that speed introduces a new gap in the train passenger's perception that becomes filled with other forms of media. LaMarre, *The Anime Machine.*
5. Akira, interview by the author, September 23, 2005, at an izakaya in Shinjuku, Tokyo.
6. Michiko, interview by the author, November 13, 2005, in Tokyo.
7. Georg Simmel famously aligns such ambivalence with an "internal reserve" that individuals develop in response to the overwhelming stimulus of metropolitan life. See Simmel, "The Metropolis and Mental Life."
8. Schivelbusch, *The Railway Journey.*
9. LaMarre, "Living between Infrastructures."
10. Yoshihara Mihoko (Senior Project Director of Planning and Production at JEKI), interview by the author, June 16, 2005, in Ebisu, Tokyo.
11. Ito, Okabe, and Matsuda, eds., *Personal, Portable, Pedestrian.*
12. Ibid, 1.
13. One of the bestselling novels in 2007 was originally a *keitai* novel composed by a twenty-one-year-old woman on her *keitai* during her commute. See Norimitsu Onishi, "Thumbs Race as Japan's Best Sellers Go Cellular," *New York Times*, Jan. 20, 2008, Asia Pacific Section. The *keitai* novel, or *keitai shosetsu*, is a form of cell-phone novel.
14. Although email sent from a computer can be received, all email from a non-*keitai* origin can be blocked.
15. Augé, *In the Metro*, 28–30.
16. Yoshihara Mihoko, interview.
17. Noda Masao, *Shōsen denshashi kōyō, taishōki tetudōshi shiryō* (Nihon keizai hyoronsha, 1991), 253, cited as well in Mito, *Teikoku hassha*, 126–27.
18. Cook, "Meanings of Non-Referential Indexes," 38.
19. Mazzarella, "Affect: What Is It Good For?"
20. Ibid., 298–99.
21. Ibid., 299–300.
22. Ibid., 299.
23. See for example Buckley, "Altered States," 347–72.
24. In "Intimate Alienation," James Fujii provides an excellent example of this in his analysis of a short story from 1907 involving a middle-age male

commuter-protagonist who exploits the proximity to women provided by the packed train to stoke his erotic fantasies.

25. See Horii, *Josei senyō sharyō no shakaigaku*; Steger, "Negotiating Gendered Space on Japanese Commuter Trains"; Freedman, "Commuting Gazes"; Tanaka, "Shanai kūkan to shintai gihō"; Adam and Horii, "Constructing Sexual Risk."

26. See for example http://chikan-g.com/schedule/. Legally speaking, the sexual-entertainment industry is not prostitution. The latter is defined as a service offering coitus for cash while the former is sexual service without coitus. Naturally, the line between the two is thin, to say the least, and the distinction is better understood as a legal ruse that allows the government to manage prostitution and receive the economic benefits from taxes.

27. Freedman, *Tokyo in Transit*, 57.

28. For a sociological perspective on the women-only train car in Japanese, see Horii, *Josei senyō sharyō no shakaigaku*.

29. It is important to note that although no female commuters with whom I spoke shared firsthand accounts of molestation, I heard numerous second-hand accounts from female foreign friends living in Japan. I thus assumed that Japanese female commuters felt uncomfortable conveying their experiences to a male researcher. In addition, without formal training in speaking with victims of sexual assault I felt it would be highly unethical and potentially harmful to victims to pursue the issue while interviewing female commuters.

30. Yuko, interview by the author, September 10, 2005, at the Kichijoji Starbucks at the entrance to Inokashira Park.

31. Mito, *Teikoku Hassha*, 121.

32. Saito, interview by the author, November 17, 2004, at Doutor Café, Toda Koen Station, Tokyo.

33. Jensen, "Multinatural Infrastructure"; Morita, "Multispecies Infrastructure."

THREE

Operation without Capacity

In December 1996, JR East, Tokyo's largest commuter-train operator, debuted a radically new train-traffic control system to manage the gap between the *daiya*. Designated an "Autonomous Decentralized Transport Operation Control System," or ATOS for short, the new system deployed advanced information technology and networked communications to transform JR East's commuter train network into a decentralized, self-governing, adaptive system. Over the course of the next two decades, JR East introduced ATOS throughout its train network in the Kanto region, and in recent years the Tokyo Metro system has adopted ATOS throughout the city's subway network.

ATOS was one of the first major projects that JR East undertook after its formation from the privatization and breakup of the Japanese National Railways (JNR) in 1987. While the new system had enormous significance for commuter-train operations, its initial impact on commuter experience was minimal. Despite grand claims from various systems engineers and JR East managers that the system would lift the network from entrenchment in bygone twentieth-century models of urban mass transportation, ATOS did not eliminate packed commuter trains or even reduce the degree of congestion within them.[1] Instead, ATOS expanded the commuter train network's margin of indeterminacy and transformed its principle modality of operation beyond capacity into operation without limits. The result is a highly resilient and emergent system in which operational irregularities are folded into regular operational order, making the packed commuter train a more sustainable practice.

Ostensibly, ATOS was the result of a fortuitous convergence of technological advances and theoretical breakthroughs in the 1990s that arrived, *just in time*, to help JR East cope with the sudden rise of commuter suicides. As I have shown in my previous work, ATOS was instrumental in decreasing the degree of disorder caused by those events.[2] This chapter builds on that prior discussion; however, it is primarily concerned with ATOS as a schema of operation for extreme infrastructure. By virtue of its capacity for accommodating the disordering force of extreme operational contingencies through self-organizing emergence, I show, ATOS offers an initial response to the demand for extreme infrastructure adequate for a world marked by increasingly severe environmental crises. What is more, ATOS appears to have again arrived just in time (as it did with the sudden increases in commuter suicides in the 1990s) to offer cities throughout the world tangible hope for meeting the challenges of current and anticipated extreme environmental events. ATOS thus embodies a techno-optimism in the face of the existential threat posed by the Anthropocene. It takes on the promise that a technological solution to the environmental crisis will materialize with miraculous timing, such that conditions in the future will be even better than they were in the past. At the same time, if ATOS's schema of self-organizing emergence makes it an exemplary realization of the kind of extreme infrastructure necessary to contain the environmental disasters of the future, it also makes it an outstanding instantiation of the contradiction whereby extreme infrastructure has become the foundation for an extreme capitalism that promises to bring about those environmental disasters. Extreme capitalism is about a new, radical level of just-in-time capitalist production and consumption without limits. It is materialized via infrastructure designed not only to endure and adapt to the dynamic and volatile conditions of a global market given to rapidly shifting trends in consumer desire but also to generate those conditions of metastable volatility. Extreme capitalism takes shape in the concatenations of high-frequency trading on Wall Street, real-time information exchange systems that connect advertising data to user impressions to generate our customized online experiences, and a global network of just-in-time commodity production, delivery, and consumption.[3] Under the logic of extreme capitalism, people like Donald Trump become icons of economic production not because of their business acumen, but merely for their ability to endlessly generate extreme media events.

The principle that "irregularity is regular" is central to the emergence of just-in-time extreme infrastructure and extreme capitalism. Adopted from theories of emergence developed at the intersection of life sciences, physics, and computer science, the principle of irregularity as regular provides the essential mechanism for technical infrastructure to accommodate both anticipated environmental catastrophes and capitalism's drive toward boundless exploitation. Mainstream technological discourse typically presents the realization of decentralized, self-organizing, emergent infrastructure as both evidence of technological advancement and a forgone necessity due to the limits of centralized structures. By contrast I argue, through the example of ATOS, that there is nothing inevitable about the adoption of the principle of emergence for infrastructure engineering. Extreme infrastructure, I maintain, is not an unfortunate but necessary precaution that has materialized just in time to help us contend with the inexorable disasters of tomorrow; it is, rather, a product of past crises and present failures. Extreme infrastructure was made thinkable when the principle of irregularity as regular became a recognizable solution, starting in the late 1960s, for global crises of capacity in labor, environment, and economy.

In exploring how extreme infrastructure became thinkable, I track ATOS's evolution from the genesis of its schema of operation in the global crises of the 1970s. In so doing, I bring forward a second concern regarding the technicity of the new system. To recall, my claim in this book is that the ethical connotations of technicity are part of what makes it a particularly powerful concept for thinking with technology. Technicity disrupts the monolithic value of the term *technology* by asking us to think in terms of "the degree of concretization" of a technical element, machine, or ensemble. Simply put, the greater the degree of concretization, the higher the technicity and capacity for collective becoming. Concretization in this regard denotes an emerging functional coherence among heterogeneous parts (and realities). It is irreducible, however, to technological optimization, which is understood in instrumental terms as the hyper-rationalization of a technological system under capitalism's drive toward market efficiency.[4] In Simondon's writing, optimization refers instead to the degree of an ensemble's margin of indeterminacy and thus its potential for further processes of *becoming with* its milieu. As in Donna Haraway's work, the concept of "becoming with" carries significant ethical gravity.[5] It is not about the fusion of one thing into another but rather about the formation of a novel articulation of generative interdependence through communication across systems (or milieus) with different orders of magnitude. For Simondon, the evolution toward technical ensembles characterized by a higher degree of technicity is a process of technical systems "becoming-organic" through their realization of a level of concretization analogous with organic individuals.[6] Accordingly, for Simondon, true technological innovation involves the expansion of a technical ensemble's margin of indeterminacy toward greater concretization and increased technicity. In Simondon's thinking, such innovation appears as an expression of a "technical mentality" informed by a desire to develop a provisional resolution for the conflicts within and limitations of a technical ensemble.[7] Thus Simondon seems to ascribe an element of technical purity to genuine engineering innovation. ATOS complicates this picture.

By virtue of its incorporation of the principle of irregularity as regular, a principle taken from living organisms, in order to realize an emergent and resilient system, ATOS embodies the ideal of becoming-organic. At the same time, ATOS exposes the underlying contradiction whereby the realization of higher technicity has become a means of adapting to climate change while enabling limitless consumption. In other words, becoming-organic has become complicit with extreme capitalism. In making this claim, my argument echoes Melinda Cooper's thesis that capitalism has learned to exploit "life as surplus" through biotechnology in ways that open new and vast frontiers of seemingly limitless capital.[8] There is considerable affinity between the conceptual genealogy behind ATOS that I present here and the critical history that Cooper offers concerning biotechnology. Both stories converge in industrial capitalism's crises of capacity circa 1970 and in the corollary reconceptualization of nature as a metastable system in which "irregularity is regular." Cooper, however, emphasizes identifying the specificity of contemporary neoliberalism through its exploitation of a novel, organic logic. Neoliberal reform in Japan is certainly part of the story of ATOS, but attributing all of the ATOS project to neoliberalism would require a reduction in complexity, particularly in terms of the motivations of the principal engineer behind its central, organic concept of "distributed autonomy." I show that insofar as the discourse regarding crises in capacity informed his efforts to develop decentralized and emergent infrastructure, he was equally inspired by the kind of speculative utopianism that drove cybernetics and is part of Simondon's writing. Consequently, the story of ATOS's emergence, presented in this chapter, resonates with the media historian Orit Halpern's exploration of Korea's new smart-city complex of Songdo. Halpern's text embarks from and ends with Songdo, yet Songdo is far from inevitable in her argument. As Halpern's work demonstrates, Songdo cannot be traced easily to radical historical shifts or neoliberalism.[9] It appears, rather, as a monstrous contradiction that elicits the questions "What went wrong?" and "How did the philosophical, post-human project of cybernetics end up producing an insatiable desire for networked environments and the dream of a limitless data economy?" Or, to put it differently, how was data transformed from a tangible aesthetic articulation of immanent utopia and thus something "beautiful" into a mundane instrument of neoliberal economic value? ATOS poses a similar question. It seeks to understand how Tokyo's commuter train network has become an exemplary instantiation of extreme infrastructure and a medium of extreme capitalism.

ATOS: Somewhere in Tokyo

Wolfgang Schivelbusch, in his history of the railroad in North America and Europe, describes the railroad as a technical ensemble that demands centralization in order to coordinate train traffic over the vast and complicated network without catastrophe.[10] A system with centralized command and control has remained to this day an idealized schema for railroad operation. In the 1920s, railroad-technology engineers in the United States developed a system called Centralized Traffic Control (CTC) that has served as the primary control technology for railroads throughout the world. Although some privately-run railroad firms in Japan adopted the CTC system prior to World War II, it did not become widespread in the country until JNR began introducing it in the 1950s.[11] To recall my discussion in chapter 1, under the CTC model a dispatcher located in the system's central control room monitors the progress of train traffic on a large schematic of the system and issues orders to trains and stations. The intensity of the dispatcher's labor depends to a great extent on traffic density along the train line as well as on the level of automation in the system. In less-automated CTC systems, the dispatcher is tasked with remotely operating switches and signals and routing the train to the appropriate platform at each station. Regardless of the degree of automation, the dispatcher's objective is to remain as faithful as possible to the principal *daiya* while allowing enough *yoyū* for operation beyond capacity.

The central command and control room for JR East's Tokyo commuter-train network is the Tokyo *shireishitsu*. Tokyo commuters know of the *shireishitsu* from the traffic updates it issues every morning for television and radio news, as well as the occasional "special traffic report," which is invariably presented by a female JR East employee. The location of the facility, however, is kept secret. Overseeing more than twelve thousand trains on eighteen different train lines for over seventeen million passengers daily, the JR East *shireishitsu* is the center of the busiest train network in the world and at the core of mobility in Tokyo.[12] As one JR East employee explained to me, "if the *shireishitsu* goes, all of Tokyo goes."[13] Although perhaps an exaggeration, the statement was meant to impress on me the seriousness of JR East's security around the facility. Indeed, the *shireishitsu* is not an easy place to get into. Only after pursuing every connection I had for a year and a half did I finally secure permission to tour the facility. In November 2005, I met up with a group of eleven new JR East employees on the day of the tour outside a train station somewhere in the city.[14] All of the new employees were male, in their early twenties, and dressed in the standard new-employee blue suit. Together with two technicians from the *shireishitsu*, who served as our guides on the tour, our little group walked a short distance from the train station to an unremarkable glass and steel building situated in the midst of multilevel highways and train tracks. At the entrance to the building, we all received a small pin bearing the official JR insignia to clip to our shirts. Our guides then led us through a pair of smoked-glass doors, into an elevator, and down the hall to a set of nondescript, heavy, gray metal doors, above which hung a varnished wood sign with blackened, engraved characters that read "Tokyo Central Command Room" (*Tokyo sogō shireishitsu*).

The sign over the door was somewhat misleading. Centralized command and control is no longer the principal mode of operation followed in the enormous oval room that serves as the *shireishitsu*. In November 2005, the facility was in the final stages of transitioning from centralized to decentralized organization with the introduction of the Autonomous Decentralized Transport Operation Control System (ATOS). Each major train line in the city has a designated ATOS terminal in the *shireishitsu* that comprises an array of consoles connected via 100 MB fiber-optic cables to every station, platform information board, signal, and switch on the train line. The terminals are arranged in a large ring around the room. Administrative workstations and a large, oval wood table that serves as a meeting point during emergencies occupy the center of the room on a slightly raised platform.

At the time that I toured the facility, only one train line, the Nanbu Line, had yet to be hooked up with an ATOS terminal. Work to upgrade the Nanbu Line with ATOS was underway, and the new computer terminals had been set up in the *shireishitsu* but not yet connected. In the meantime, the Nanbu Line, which runs between Tachikawa and Kawasaki, was still operating according to a simple, centralized model requiring a dispatcher at a small table to wear a headset plugged into one of several transmitters stacked next to a phone and fax machine. A few schedule corrections appeared hastily written in white chalk beneath a row of station names on a large blackboard standing beside the table. The dispatcher was confined to his position and attentive in his task of monitoring traffic-status reports, issuing commands, and dispensing schedule changes. The setup was crude, even by CTC standards, and thus not necessarily an accurate picture of how more-complex train

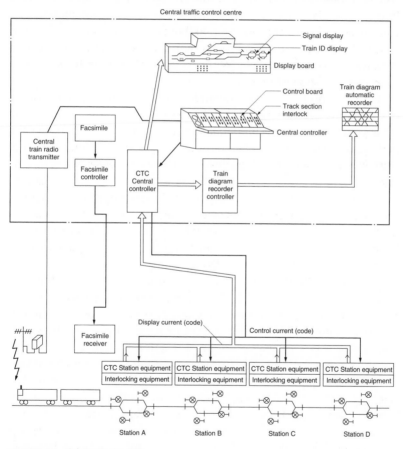

FIGURE 3.1. Schematic for CTC

Source: Takashige, Tetsuo. "Signalling Systems for Safe Railway Transport." *Japan Railway & Transport Review* (September 1999): 44–50.

lines were controlled prior to the introduction of ATOS. On the CTC for the Chūō Line, for example, the dispatcher would have had a real-time display of the train progress and a console to remotely control switches on tracks. Platform routing for stations would have also been automated through computers located either in the central command center or at stations along the line.

In comparison with the circumstances at the Nanbu Line central control station, the scene around the ATOS terminals in the *shireishi-tsu* was relaxed. At the Saikyō Line, a group of young technicians had pushed their high-back executive-style office chairs away from the computer consoles and into a circle, where they sat passing the time in conversation. The atmosphere around ATOS terminals for other train lines was similar. In most cases, only one technician was present at a console designed for three or four, looking over a printout of the *daiya* or seated before one of the computer monitors clicking through various displays of the system status. With ATOS monitoring train operation, close human supervision is unnecessary.

ATOS decentralizes command and control. This means that instead of depending on a central dispatcher to monitor traffic progress and issue commands, each component of the system is responsible for making its own decisions. The term for this system of organization is "distributed autonomy" (*jiritsu bunsan*). *Distributed autonomy* is a complicated term in Japanese, in part because "autonomy" (*jiritsu*) can be written using two different homophonic character combinations, only one of which is correct. The incorrect combination uses the characters for "oneself" (*ji*) and "stand" (*ritsu, tatsu*) to mean "independence." In the correct spelling, the character for "oneself" (*ji*) is combined with the character for "rhythm, law, regulation, control" to mean "self-control" in accordance with the philosophical and juridical notion of autonomy. Autonomy or "self-control" in this regard is not the capacity to act independently. It denotes, rather, the capacity to act through interaction via communication in order to remain in accordance with others who constitute the field of one's possibility for action. This principle of communicative interaction is the key facet of decentralization under ATOS.

ATOS, it is important to emphasize, does not transform the underlying principle of traffic control via regulating the gap between the principal (*kihon*) and the operational (*jisshi*) *daiya*. Rather, it changes the organizational schema of the gap and, consequently, how the gap is conceptualized. In the centralized system under CTC technology, order is imposed from the top down. In the decentralized system under ATOS, order is allowed to emerge from the bottom up. In a paper

情報化時代にふさわしい
先進の輸送管理業務を実現します

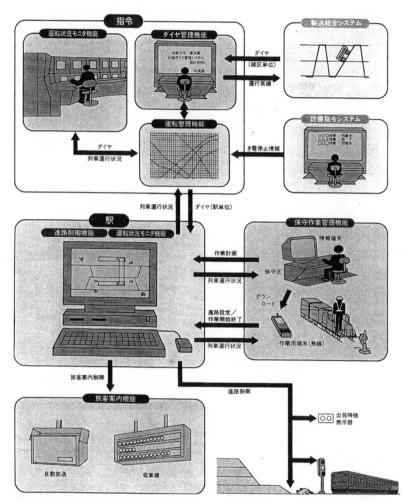

FIGURE 3.2. Schematic for ATOS showing the flow of information between different components of the system. (Japanese reads: "An Advanced Transport Management Befitting an Information Era.")
Source: *Japanese Railways Engineering Journal*, no. 135 (1995).

delivered at the International Symposium on Autonomous Decentralized Systems (ADS) in the early 1990s, two Japanese engineers summarize the implications of the latter for realizing self-organizing emergence through collective autonomy: "The autonomous decentralized systems are systems in which the functional order of the entire system is generated by the cooperative interactions among its subsystems, each of which has the autonomy to control a part of the system."[15] As the diagram above suggests, how that interaction actually transpires at the level of software and hardware is immensely complicated, although easy enough to parse in a schematic fashion. Basically, train stations act as nodes in the system, collecting information that is then shared with other system components in a general "data field." ATOS does not alter the general topography of the network. Stations, tracks, switches, and signals stay where they were under centralized control. Instead, ATOS transforms the mode of interaction between each system and subsystem. The result is a new topology of bottom-up emergent organization.

For the *shireishitsu* technicians, the difference between CTC and ATOS is articulated as a shift from managing (*kanri*) to supervising (*kanshi*) the gap between the *daiya*. Whereas managing involves active logistical attention to and administration of the order and flow of trains (exemplified by the dispatcher for the Nanbu Line), supervising is more passive, requiring only observation of the unfolding order. One tangible expression of the shift from managing to supervising is the disappearance of the display board, the large schematic of the system that allows dispatchers to follow train movement, from the central command room. With ATOS, the display board is replaced by the emergent pattern of the system, which the *shireishitsu* technicians can watch, if they want, as an emerging *daiya* on one of the ATOS monitors. (The appearance of lines and numbers on the screen recalls how the emergent order of the simulated world is represented as a cascading green pattern on a computer in the Wachowskis' 1999 hit science-fiction film *The Matrix*.) As a visual medium of command and control that places the dispatcher in a privileged Archimedean point over the network, the display board embodies a panoptic desire for totalizing vision and knowledge. In his famous reading of Jeremy Bentham's panopticon prison, Michel Foucault describes the panopticon as an idealized diagram of power "abstracted from any obstacle, resistance, or friction" of actual collective life.[16] In Foucault's analysis, the panopticon iterates a modern disciplinary logic of centralized power from which all irregularity and uncertainty must be expelled. In later work, Foucault amends his explication of modern power with his thesis on security.[17] Whereas discipline works on individuals in modular spaces of discipline's own making, a regime of security follows a decentralized schema and operates on populations through existing networks of circulation in order to modulate, rather than banish, irregularity and uncertainty. Although Foucault did not intend it, his schematization of the transition from discipline to security maps onto a narrative of a historical shift from a modern industrial economy to a postmodern, postindustrial information society, and by extension maps onto the notion of a shift from twentieth-century capitalism to twenty-first-century neoliberalism.

The shift from managing the *daiya* under CTC to supervising it via ATOS complicates this historical narrative. Even with the centralized system under CTC, insofar as the principal *daiya* represented an ideal order, divergence from it was not strictly the result of "obstacle, resistance, or friction" but rather a necessary factor of operation beyond capacity. As I argued in chapter 1, if train operators had insisted on strict adherence to the principal *daiya*, the system would never have been able to accommodate the city's commuter population. Divergence was a matter of finessing the system's intervals for operation beyond capacity. The move from a centralized (CTC) to decentralized (ATOS) system is thus irreducible to a historical shift from disciplinary structures requiring absolute compliance to norms to a processual regime of modulated irregularity. With both systems, rather, the central matter is the dynamic between structure and process, control and emergence. They differ primarily in how they negotiate this dynamic. In the centralized system, the dynamic is arbitrated by managing the different actors within the network, whereas in the decentralized system, under ATOS, operators supervise the scene of interaction, giving them a more direct access to the network's margin of indeterminacy. This contrast lends itself to a different conceptualization of the gap between the *daiya*. Under a centralized schema, the gap is an inevitable but containable force of entropy, whereas ATOS, as I will argue, allows for thinking of the gap as a necessary condition of emergence.

What is more, in shifting the focus of operation from the network actors to the space of interaction, ATOS allows for greater centralization of command than the centralized system ever did. As I mentioned in chapter 1, under the CTC system, the complexity and density of traffic on main train lines prevented train companies from implementing absolute centralized control. Train traffic simply generated far too much information for a single controller to manage and respond to in a timely manner. As a result, it was common for main train lines to have a number of control centers that would then report to the central controllers in the *shireishitsu*. Despite its underlying decentralized schema, ATOS allows for instant and total recentralization. Such recentralization occurred twice during my tour of the *shireishitsu* with the sounding of an alarm bell that brought technicians quickly back to their posts to take control over the system. In both cases, the alarms involved minor incidents and demanded only slight revisions to the *daiya*, which were made on a computer via a click of a mouse after a brief consultation with the relevant train station by phone. Once the revisions were made, decentralized command was handed back to ATOS.

The Capacity Problem

When the media theorists Alexander Galloway and Eugene Thacker riff on Deleuze to declare "we are tired of rhizomes," they draw attention to the appropriation of distributed topology as a techno-organizational medium of neoliberal biopolitics and its subsequent collapse as a schema of resistance.[18] Galloway and Thacker don't so much critique Deleuze as criticize the tendency in popular (and, to a certain extent, academic) media discourse to equate technological topologies with sociopolitical forms of organization.[19] According to this discourse, mass-mediated society inevitably results from centralized systems that relegate individuals to objects of communication and control, while decentralized networks produce autonomous individuals within a self-organizing, emergent collective that gives rise to such (potentially) emancipatory techno-social forms as peer-to-peer culture, flash mobs, and Facebook. It is not difficult to imagine how ATOS, with its principle of distributed autonomy, could be perceived as potentially liberating commuters from twentieth-century modes of urban mass transportation.

But ATOS was never intended to emancipate commuters from the packed train. It was meant rather to mitigate the effect that extreme events in its operational environment have on the system, and in so doing transform operation beyond capacity into operation without capacity. As is often the case with technological innovations, a plethora of articles in scientific and railroad journals, many of them presented at biannual international workshops and symposia on Autonomous Decentralized Systems (ADS) that have been held since 1993, preceded the introduction of ATOS.[20] Without exception, the early literature from these meetings is devoted to enumerating the shortcomings of centralized traffic-control systems while praising the merits of distributed autonomy for bringing railway management into the twenty-first century.[21] The underlying flaw of centralized systems, they argue, is that they are labor-intensive. What they mean by labor is capacious; it can refer to the pressure on central dispatchers to maintain the integrity of the gap, the difficulty of *daiya* recovery in times of disorder, the necessity to convey information updates to commuters, and system maintenance, among other things. All of these labor issues become expressed as a single defect, which is that centralized systems lack the flexibility to handle irregularity without an enormous influx of manpower. For Tokyo's commuter-train operators, this lack of flexibility manifested particularly under two different but related circumstances: during sudden surges in commuter demand and during instances of disorder in the *daiya* as a result of technological malfunctions or commuter suicide. Both circumstances constituted extreme events in the system's operational environment, and both could cause massive systemic failure.

Daiya technicians (*sujiya*) have long aspired to make the margin of indeterminacy a legible object of logistical management and prediction. From the early decades of commuter trains in Japan, they learned to factor in anticipated increases in transport demand around such things as sporting events, concerts, national holidays, and so on into their calculations. But they could not anticipate every contingency. A political event such as a street demonstration, an unusually important sports game, or even a consumer event in the city could cause an unforeseen rush of commuters on a train line, creating disorder that threatened operators' capacity to manage the gap between the *daiya*. If such extreme events were barely manageable, a technological malfunction or a suicide on a train line during rush hour sent commuters pouring into nearby stations in search of alternative routes, putting the system in jeopardy. Even with a certain number of trains always on standby for such contingencies, train operators could not stream trains to stations fast enough to accommodate the demand with such short notice. Platforms would become even more crowded, increasing the likelihood of a commuter falling on or being pushed to the tracks in front of an arriving train. In sum, such instances of unanticipated surge in commuter demand created a disorder that metastasized throughout the system, engendering the collapse of the principal *daiya*. In response, *daiya* technicians in the central command room would have to hastily calculate and draw up a provisional *daiya*. Once calculated and drawn, the provisional *daiya* still had to be transmitted by fax machine to each station in a network, one at a time, before order could be restored and train service resumed.

Throughout the postwar period and the 1980s, train companies managed to mitigate instances of unpredictable disorder by maintaining an enormous labor force. Mito Yuko, in her history of Japan's railroad, tells how, under JNR, teams of university students were employed as part-time workers on call, ready to rush to city stations in the event of extreme operational irregularities in order to help manage swells in platform crowds.[22] Why and how unpredictability became unresolvable through labor is related to shifts in the political economy that began in the 1970s, to which I will return later in the chapter. For now, however, I want to explore the significance of operation beyond capacity through ATOS.

Operation without Capacity

By mobilizing the organizational schema of distributed autonomy, ATOS transformed JR East's Tokyo commuter-train network from a system encumbered by a fixed operational limit into a system able to accommodate extreme environmental events as manageable irregularities within the regular operational order. The result was a system with a highly flexible peak-performance threshold. What does this transformation look like? The information-science scholar Yamamoto Masahito offers a helpful analogy that compares the old centralized train-traffic system (CTC) to a conventional sit-down sushi restaurant and ATOS to a conveyor-belt sushi restaurant (*kaitenzushi*).[23] At the latter restaurant, customers sit at a counter fitted with a conveyor belt and select the sushi they want from among the color-coded plates circulating on the conveyor belt. The color coding indicates price, and customers are charged according to the plates they have gathered at the meal's end.

The latter arrangement allows for operation without thresholds. In a sit-down sushi restaurant, Yamamoto explains, the sushi chef begins preparing the sushi only after receiving an order from the customer. Similarly, in the centralized traffic-control system, the dispatcher waits for an order from stations before sending trains to that station. Both the sit-down sushi restaurant and CTC thus have a clear operational threshold. If the sit-down sushi restaurant is suddenly flooded with customers who all want their orders filled right away, the sushi chef becomes overwhelmed in the same way that the dispatcher becomes overwhelmed by a sudden surge in commuter demand. By contrast, in a conveyor-belt sushi restaurant, the sushi chefs do not wait for orders from the customer.[24] Instead, the chefs determine on their own what kinds of sushi to prepare based on what they discern from watching the conveyor belt. The conveyor belt, suggests Yamamoto, is analogous with the "data field" in ATOS. It is a place where the customer and sushi chef (subsystem) can share information, making cooperative data processing possible. Because the sushi chefs in the conveyor-belt sushi paradigm do not wait for orders, but make the sushi according to their autonomous judgment, the system is not restricted by the operational limits of the conventional centralized/sit-down sushi restaurant. Although Yamamoto presents the operational peak of the conveyor-belt paradigm in a diagram curve that only extends somewhat above the threshold of the sit-down restaurant model, he writes in the body of his argument that with the conveyor-belt sushi restaurant, "no peak exists" (*pīku sonzai suru koto naku*).

In reorganizing the commuter train network, ATOS shifts the network's operative emphasis to enabling an emergent process in which irregularity in the form of extreme environmental events is enfolded into the regular operational order. In so doing, ATOS provides far more than a more efficient traffic-management system: it offers a preliminary instan-

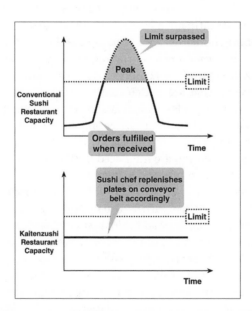

FIGURE 3.3. Sushi-restaurant schematic. Top chart shows the conventional sushi restaurant as a system that quickly passes its operational threshold. Bottom chart shows the conveyor-belt sushi restaurant where production remains absolutely steady and safely below the limit. Source: Yamamoto, Masahito. "Sekai ni hirogaru jiritsu bunsan." *Landfall* 48 (2003): 1–5.

tiation of emergence by design on a massive infrastructural level. In the scientific and corporate literature that accompanied the development and introduction of ATOS, the need for the new system was presented as both inevitable and intuitive. As will be seen below, however, the emergence of ATOS was neither inexorable nor certain.

Extreme Infrastructure and the Capacity Crisis

As climate change becomes unstoppable, the threat of extreme environmental events renders ATOS's schema of operation not just an ideal but an imperative for infrastructure in the twenty-first century. Consider for example the statement that greets visitors to the United States National Climate Assessment Report website: "Infrastructure is being damaged by sea level rise, heavy downpours, and extreme heat; damages are projected to increase with continued climate change."[25] Produced by GlobalChange.gov, the website mobilizes a combination of "key message" statements over full-screen images of heat-warped roads, flooded subways, and washed-out bridges to summarize the findings of the three hundred experts who contributed to the National Climate Assessment Report. Its ultimate message is that the worst has yet to come. Already crumbling under decades of neglect and mismanagement, the nation's outdated but vital transportation, urban water, and power infrastructures will certainly succumb to the onslaught of extreme environmental events that are anticipated to devastate cities across the globe in the coming decades. The dismal forecast for the future of national and global infrastructure is offset only by the fluid interface and stunning images on the website, in which scenes of projected devastation and rampant pollution are interlaced in sliding frames with idyllic panoramas of nature and interactive graphs.

The Anthropocene has created a new kind of capacity problem—there is simply too much weather. Urban power grids are collapsing under the stress of millions of additional air conditioners clicked on to alleviate the effects of surging summer heat waves; sewer and water systems designed in accordance with weather data from the past two centuries cannot handle the onslaught of unprecedented torrential rains; and transportation, processing, and tracking systems cannot respond quickly enough to stem the tides of refugees pouring across borders seeking relief from famine-ravished and war-torn lands. With forecasts for the future taking increasingly apocalyptic tones, self-organizing emergent systems have become the ambition of governments, urban planners, and corporations throughout the world in cities eager to transform crucial infrastructure into resilient ensembles able to withstand increasingly severe environmental events.

ATOS's timing appears impeccable. Distributed autonomy offers a schema of operation that seems tailor made for the capacity challenges of extreme environmental events in the present and future. As such, ATOS comes to embody the hope of technological salvation from the imminent threat of ecological collapse. It becomes the promise that through humanity's technological ingenuity, the pattern of technological development will continue and our children will inherit an even better world than we inherited from our parents. The problem is that ATOS does not exactly represent humanity's inexhaustible potential for last-minute innovation to thwart impending disasters. As I show below, the origin of its schema of operation lies not in the anticipation of an apocalyptic future but rather in past crises of capacity. Consequently, ATOS suggests not hope but rather a troubling feedback loop between the past technological finesses for provisional resolutions of conflict and the (seemingly) ineluctable innovations of the present and future.

The Centralization Paradigm and Capacity

Stephen Graham and Simon Marvin, in their seminal history of urban infrastructure, track the rise and fall of the "modern infrastructural ideal," from 1860 to 1960, as the dominant Western paradigm for urban development.[26] The modern infrastructural ideal elevated rational urban design to a moral imperative. It was the implementation of urban planning in its highest form, giving rise to Georges-Eugène Haussmann's Paris and Robert Moses's New York in conjunction with a high modernism that saw the pursuit of science and technology as a means of bringing a civilizing order to the perceived disorderly forces of nature. Under the modern infrastructural ideal, built systems were deployed to engineer the social order. More importantly, under the modern infrastructural ideal, industrial capitalism surged. Cities expanded at fantastic rates, factories proliferated, and populations swelled, while rivers and lakes filled with toxins, forests were cleared for timber or space, and the air became a miasmic concoction of noxious gases and hydrocarbons.

As a design paradigm, the modern infrastructural ideal peaked sometime in the early twentieth century. But it did not unravel entirely until the late-1960s culmination of economic, environmental, and social crises in industrial capitalism.[27] To a great extent, its fall was an international phenomenon marked by similar crises erupting in Chicago, New York, Tokyo, and Paris. While the particulars of these crises differed, their origin lay in what could be called a general-capacity problem. Industrial production had reached a capacity threshold on economic and environmental fronts. While a growing decline in demand (industrial production had made too much) threatened industries with lost revenue, concern emerged about the dismal state of delicate natural ecologies that were collapsing under decades of pollution. For many social, political, and economic thinkers of the time, the combination of crises marked a clear limit to industrial capitalism as a sustainable political, economic, and social system. Such sentiment received its clearest articulation in a report titled *The Limits to Growth*, published in 1972 by a think tank called the Club of Rome.[28] Authored by four individuals with backgrounds in environmental science, education, and climatology, the report drew on a long-running conversation with scholars from a wide range of disciplines in ten countries, including the head of the Economic Research Center in Tokyo, to become a galvanizing argument for environmental movements around the world.[29] Overall, the document presented a very grim picture of the future ecological health of the planet and the well-being of populations under industrial capitalism. Exploiting new computer technology for modeling complex, nonlinear dynamic systems, it predicted that if population growth, use of natural resources, and pollution were not curtailed to ecologically sustainable levels, the global ecology would collapse, producing a catastrophic effect on human life and society. In other words, operation beyond capacity under industrial capitalism had reached a limit.

The capacity crises in Japan reflected much of what was happening elsewhere in the world at the time. Japanese historian Tessa Morris-Suzuki, in outlining the specificity of the nation's crises in the 1960s, points to four factors: a decrease in the rate of urbanization, a falling birth rate, rising costs in labor, and increasing foreign-trade barriers against Japanese imports.[30] In addition, she argues, the Japanese public was beginning to express concern over the effects of industrial pollution through a "rejection of economism," especially in the wake of several cases of environmental damage in which many of the victims were babies and children. The most infamous case involved mercury poisoning from industrial dumping in and around the harbor town of Minamata on Japan's southern island of Kyushu.[31] According to Morris-Suzuki, the rejection of economism also involved shifting sentiments regarding labor and employment among the younger generation. As

the generation of postwar baby boomers—often called "the generation that grew up without war" (*senso o shiranai kodomotachi*)—moved
through the universities and prepared to enter the labor force, they appeared less willing to accept the tropes of devotion to the company
and individual sacrifice that had secured the commitment of labor in
the first phase of Japan's postwar economic recovery. Their frustrations
with the growing inequity and rigidity of the system, writes Morris-
Suzuki, contributed momentum to the student demonstrations opposing the Japanese government's signing of a security treaty with the
United States (and later the demonstrations opposing the war in Viet-
nam) that swept through the universities and streets of Tokyo in the
late 1960s.[32]

When the oil shock hit in 1973, Morris-Suzuki suggests, it merely
precipitated a final unraveling of an already unstable constellation.
But it also created the sense of a real "mode of crisis."[33] As in other
countries, in Japan the sense of crisis provided the opportunity that
pro-monetarist (often glossed as neoliberal) economic-policy mak-
ers, influenced by Milton Friedman, had been waiting for to institute measures for dismantling the nation's postwar labor and industry structure.[34] JNR became a primary target. With its large workforce,
JNR was home to three powerful labor unions. The largest among the
three was Kokurō, which had close to half a million members at its
peak and maintained strong ties with the Japan Socialist Party (JSP).[35]
Pro-monetarist economists knew that JNR was the keystone in Japan's
postwar national-industry structure. If it could be taken down and its
unions broken up, a path would open for the reorganization of other
major national industries such as Nippon Telegraph and Telephone
(NTT) and the Japan Postal Service Agency.[36]

JNR was a profitable enterprise until 1964. Its financial troubles began in part with the growth of the nation's automobile and airplane industries and the subsequent decrease in rail customers. But these financial troubles were equally the result of government pork-barrel projects
that exploited the industry for the sake of rural constituents. Nevertheless, labor was made to bear most of the blame, with the claim that the
industry had too many employees—operating beyond capacity, as it
were. Labor unions fought back. Between 1964 and 1980, JNR management and the Japanese government's Ministry of Transport made six
separate unsuccessful bids to decrease necessary labor by rationalizing
JNR operations.[37] Tensions between the railroad labor unions and management reached a peak with a 1970 government and JNR initiative
called the "Productivity Improvement Program" (*maru-sei undō*). The

program, which was perceived as a threat by Kokurō, ultimately back-fired, leaving the leaders of Kokurō in a position to dictate operating conditions and exact revenge on anyone who had sided against them.

Although the public tended to side with the railroad labor unions in these confrontations, public sentiment shifted clearly in 1975 when Kokurō embarked on an eight-day strike for the right to strike, with the aim of bringing the economy to its knees. As Charles Weathers writes concerning the shift in public support at that time, "television footage of office workers trudging miles to work is well remembered in Japan."[38] The public's opinion was impacted by more than the tremendous inconvenience to commuters from the strike: Japanese society at the time had grown exhausted by leftist labor politics. Street demonstrations in Tokyo in the late 1960s, led by leftist student movements opposing Japan's re-signing of a military cooperation treaty with the United States, among other things, had become violent and captured the nation's attention only to conclude without tangible results. In addition, the national broadcast of the Asama-Sansō incident in 1972—in which a group of United Red Army members took the wife of a lodge keeper in Nagano as a hostage and held off the police for nine days while pursuing a bloody purge of their own members—did much to further tarnish the public impression of leftist movements as sense-lessly violent and dogmatic.

As a result, the Kokurō eight-day strike for the right to strike proved a terrible miscalculation that had terminal consequences, as it was instrumental in helping Nakasone Yasuhiro become prime minister in 1982. Known for his pro-monetarist policies and hostility toward labor unions, Nakasone immediately set to work on privatizing and breaking up JNR. After leading his Liberal Democratic Party to a decisive victory in the 1986 general elections, he pushed the final legislation to break up JNR through the Japanese Diet. In 1987, JNR was privatized and divided into six regionally specific passenger-rail companies, a research organization (RTRI), and an information-systems company (JR System). Although railroad labor unions had managed to secure a promise during privatization negotiations that redundant personnel would be retrained for positions in the new JR Group, things turned out quite differently. Retrainees were sent to the euphemistically named "Human Resources Development Centers" (*jinzai katsuyō sentā*), where they were assigned to demeaning tasks with utterly no relevance to train operation—such as pulling weeds, cleaning toilets, and painting fences—to pressure them to quit. According to a survey presented in July of 1986, among the employees reassigned to "Human Resources

Development Centers," 75 percent were former affiliates of Kokurō, the labor union that had resisted rationalization, while 22 percent were from the other two railroads combined.[39]

Technicity and Economy

When JR East inherited its large piece of Tokyo's commuter train network from JNR in 1987, it faced a new kind of capacity challenge. Privatization left the young company with the prospect of a greatly reduced labor force that would be unable to contend with unanticipated surges in commuter demand in the coming years. That such a radically novel system as distributed autonomy was less than a decade away from being ready for deployment was hardly a coincidence. Like JR East, the conceptual schema for distributed autonomy emerged in a moment of infrastructure shutdown during the 1970s' crises of capacity, albeit not in Japan. Specifically, the schema was born at 8:37 p.m. on the evening of July 13, 1977, when a series of fluke lightning strikes destroyed critical components of New York City's power grid. The damage set off a chain reaction of technological failures and human error that plunged a large part of the city into a twenty-four-hour blackout. With the city already struggling under an aging infrastructure, a rising crime rate, and fiscal hardship—imposed, in part, as a punitive measure by investment bankers—the blackout could not have come at a worse time. It is typically remembered for the looting and general social unrest it released.[40]

For the Japanese engineer Mori Kinji, the blackout of 1977 was a eureka moment that led to his formulation of distributed autonomy. Mori was in the United States at the time on a postdoctoral research fellowship at Berkeley University as a new employee for Hitachi Corporation's System Development Laboratory. He has since become something of a celebrity among Japanese systems engineers, with the story of how he developed the principles and software for Autonomous Decentralized Systems (ADS) appearing in numerous interviews and articles in Japanese technology journals. ADS has become the foundational architecture for complex multi-agent systems used throughout the world in everything from oil refineries to electronic payment technology, communication networks to product tracking. Mori's reputation as an innovator in systems engineering and a social visionary eager to move past conventional modes of thinking has earned him visiting positions at numerous universities across Japan as well as China.[41] He recently took a position heading a laboratory in Waseda University's new Green Computing Systems Research and Development Center, which is where I visited him on a warm afternoon in the spring of 2012.

When I met with Mori, he had just taken occupancy of his new office and had yet to move in all his books. The only aesthetic addition to the crisp white walls, metal cabinets, and white bookshelf with glass doors was a large poster for ATOS taped to the front of a desk at the entrance, which Mori stopped to point out to me when we entered the room. As part of an initial campaign aimed at advertising the introduction of the system, the poster illustrated the technology's underlying schema of distributed autonomy through images of the *shireishitsu*, trains, passenger information board, and signal traffic lights arranged in circles as separate systems connected to one another via two-way arrows. The emphasis is on the realization of a network of autonomous but interconnected and interdependent systems.

Mori is a slender and energetic man, quick to smile, and eager to tell the story of how he arrived at the concept of distributed autonomy, which he did for me while shifting comfortably back and forth between Japanese and English over tea and dinner at the faculty club. Mori was of course not the first to imagine decentralized systems. He was, however, the first to develop software capable of realizing a decentralized system through distributed autonomy on a massive urban infrastructural scale. His account of his innovation brought together the dilemma of contradiction that has befallen the ideal of becoming organic. On the one hand, it rehearsed a conventional digital discourse in which decentralization is envisioned as part of the inexorable forward march of technological progress necessary for bringing twentieth-century urban infrastructure in line with an emerging twenty-first-century global economic reality. In this regard, the New York blackout of 1977 carries purely economic implications, demonstrating the need for dynamic infrastructure adequate to the increasing intensity of an emerging global information economy. The rapid tempo and fluid, unpredictable modality of the emerging information-based world of the 1970s, explained Mori, demanded from urban infrastructural systems an unprecedented level of quick adaptation to changing circumstances while also requiring a high level of reliability and maintenance. Technical systems were suddenly required to stay online 24/7 in order to keep pace with an increasingly dense and connected global economy. At the same time, they needed to be able to accommodate rapid modifications in accordance with shifts in market forces while remaining resilient to malfunction. Parsing the circumstances in engineering terms, Mori defined the challenge of the time as creating technical systems with a high degree of "fault tolerance." Fault tolerance refers to the ability of a system to remain in operation with only slightly diminished performance levels despite the failure of key subsystems. It is thus a metric for a system's ability to handle irregularity.

On the other hand, in other parts of his account, Mori described his search for an innovation as informed by a philosophical concern with questions of technics. His thinking, in this regard, reflected the kind of utopian vision of a dialogic relationship of becoming with technology that spurred other projects of the time aiming to realize technological articulations of complexity. The New York blackout, in this iteration of his account, figured as not merely an economic paradox, but a philosophical one, in which decentralization became a matter of expanding a technical system's margin of indeterminacy. As Mori put it, the blackout demonstrated a conceptual impasse at work in the field of engineering. The failure of engineering to develop decentralized systems at the time, he contended, was due not to a problem of technological capability but to a failure in thinking. Namely, engineering's commitment to the centralized model was the consequence of an epistemological intransigence and corollary methodological investment in the field to an "absolute perspective" (*zettai shiten*) that accepted only one view of reality and truth. Adherence to absolute perspective, Mori believed, was the effect of the reductionist approach taken in modern science, which sought to distill observable phenomena to their absolute underlying principles in order to produce generalized laws of nature. Transposed into a method of engineering, Mori asserted, absolute perspective led to an insistence on absolute design, dictating that the specifications of any system had to be entirely determined prior to its construction. The result was that, once built, a system could not be modified without great disruption. To put it in different terms, the result was a system with a narrow margin of indeterminacy and a corollary limited technicity. Whereas Mori had earlier introduced the issue of infrastructural rigidity as an economic dilemma, he reintroduced it as an ontogenetic problem in respect to the problem of absolute perspective. Absolute perspective, he contended, produced a technologically determined reality that forced human behavior to conform to mechanistically fixed operational thresholds. In so doing, it shut down avenues of becoming.

Offering an analogy, he pointed to the previous generation of electronic ticket gates at train stations, which required commuters to insert an actual paper ticket that the machine would then process and discharge on the other side for the commuter to collect. The ticket gate was limited by operational thresholds to accept only a certain number of commuters per minute, explained Mori.[42] If too many commuters tried to pass through the gate in one minute, the system's computers could not keep pace with the calculations and would shut down, causing the gate to close. To overcome this kind of technological restriction, Mori sought to realize a form of "relative perspective" (sōtaiteki shiten) in technological systems. Achieving relative perspective through a technical ensemble held the promise of liberating technology from the restraints of modern science. Or, as Mori put it, it would engender a historical shift from "an age dominated by science to an era of technology" (saiensu no sekai kara tekunorojī no jidai e). To explain what he meant by the latter, he returned to his ticket-gate analogy and the new generation of ticket gates, which use distributed-autonomy architecture to overcome the operational limits of the older machines. I will say more about these devices below; in the meantime, suffice it to say that commuters no longer have to insert a paper ticket. Instead, they need only touch a so-called smart card with an embedded Integrated Circuit (IC) to a sensor on the gate. An initial fee is subtracted from the card at the entrance. When they exit, the final fare is calculated, based on distance traveled, and subtracted from the card. For Mori, the new ticket gates instantiate relative perspective by not requiring commuters to adjust their pace in accordance with the ticket gate's processing thresholds. The electronic ticket gate, rather, conforms to human behavior. If it is unable to calculate a commuter's fare fast enough, it tags the smart card so as to be able to catch up with the commuter and finish the calculation later in the system. In other words, like ATOS, the new ticket-gate system uses distributed autonomy to handle irregularities.

For Mori, the new ticket gate gestures to something much more than overcoming a capacity problem. Its significance lies in its potential to liberate commuters from the dictates of technological thresholds. Moving seamlessly from technological potential to questions of governance, Mori contended that distributed autonomy will transform Japan's managed society (kanri shakai) into a society of emergence and, in so doing, end the stagnation that social critics see as having stifled innovation while trapping the nation in recession. Japan's population harbors a rich potential for innovation in all fields, claims Mori, but it is being suppressed by a rigid and bygone sociopolitical structure that favors precise planning and massive centrally administered organizations. For Mori, Japan's reliance on nuclear power represented the most egregious example of this structure. The previous generation of electronic ticket gates was, it seems, also a metaphor for technologies of governance for Mori.

Although Mori was not aware of it at the time, his pursuit of a technological means for realizing relative perspective put him in sync with contemporary developments in theoretical physics, in which advances in computer technology were allowing for modeling relativity and fueling a turn to complex dynamic systems theory. According to James Gleick, absolute perspective (otherwise known as Newtonian determinism) was indeed for a long time the driving principle behind modern physical science, compelling scientists to "break their universes down into the simplest atoms that will obey scientific rules."[43] That approach informed initial developers of the computer and computer science, such as John von Neumann, who is said to have argued adamantly that it was unnecessary to acknowledge relative variables in forecasting events within open systems.[44] Complexity theory, by contrast, endeavors to embrace relative perspective by exploring the indeterminate structure and events of nonlinear dynamic systems.

In a nonlinear dynamic system, there is no center and no overarching principle for system design, which means no overarching perspective or reality-imposing order or structure over matter. What nonlinear dynamic systems do is produce self-organizing complexity.[45] Self-organization premises the emergence of order "in the absence of design" through a bottom-up process in which a multiplicity of principles, perspectives, and realities begin to organize into increasingly complex patterns of intelligible behavior.[46] The system thus displays "self-organizing" properties whereby order *emerges* from the cooperative interaction of its different parts. Complexity theory emphasizes dynamic becoming over being, as well as metastability over equilibrium, as the condition of nature and the natural condition of matter. At some point in the mid-1970s, these principles migrated into the life sciences, blurring the boundaries between living and nonliving and redefining understandings of nature and life. Attention shifted away from an association of life with sentience to an understanding of life as a process of complex emergence whereby matter self-organizes into increasingly dense networks of interaction.[47] Mobility, resilience, and a corollary capacity for adaptation became the defining criteria for life, thus displacing a previous focus on cognitive problem-solving skills and social interaction.

Research into Artificial Life (ALife) also embarked in 1987 from this conceptual genealogy of life as a property of self-organizing emergence at a conference organized by Christopher Langton and hosted by the Santa Fe Institute.[48] ALife researchers claimed that cybernetics failed to develop artificial intelligence (AI) precisely because of its reductionist approach to human perception of reality. ALife researchers instead took the emergent approach to realize models capable of self-organization within a virtual computer environment.

Finally, while Mori was developing distributed autonomy, on the other side of the world MIT roboticist-cum-entrepreneur Rodney Brooks was building his "creatures": autonomously mobile robots that exploited a schema of distributed autonomous functions that he called "subsumption architecture."[49] In his work, Brooks described the remarkably complex tasks his "creatures" could accomplish in the real world, at a time when mainstream robotics using a centralized processing system struggled to develop machines that could negotiate simple obstacles in a simulated, uniform environment. One of the first creatures that Brooks built, for example, was tasked with wandering around the office area of the laboratory to find, retrieve, and deposit empty soda cans in the can repository. As simple as this may sound, when broken down into constituent steps, the complexity of the operation becomes clear. The creature had to be capable of

avoiding objects, following walls, recognizing doorways and going through them, aligning on learned landmarks, heading in a homeward direction, learning homeward bearings at landmarks and following them, locating table-like objects, approaching such objects, scanning table tops for cylindrical objects of roughly the height of a soda can, serving the manipulator arm, moving the hand above sensed objects, using the hand sensor to look for objects of soda can size sticking up from a background, grasping objects if they are light enough, and depositing objects.[50]

As with distributed autonomy, Brooks's creature's abilities resulted from a functional order that emerged from the cooperative interaction of its semiautonomous components rather than from a directive issued from a centralized control mechanism. In fact, Brooks went one step further than Mori and eliminated the central processor entirely: his creatures comprised semiautonomous layers, each with a specified function that turned on or off depending on the status of a message circulating among the layers.

Using the distributed-systems principle, both Brooks and Mori produced machines that in many ways fulfilled the specification of life within the ALife project. Yet both departed from ALife in their insistence that their machines inhabit and evolve within the same complex environment as human beings rather than within virtual worlds. This meant creating machines that could cope in a timely manner with changes in a dynamic environment, possessed a high degree of fault tolerance to cope with irregularity, were able to reprioritize multiple objectives according to fluid operational conditions, and, finally, could actually do something in the world—or as Brooks says, have a "purpose in being."[51] In developing their innovative technical ensembles using distributed systems, Brooks and Mori found inspiration from similar sources in the life sciences and philosophy.

Making Irregularity Regular

When Mori was looking for an alternative to centralized infrastructure, he might have drawn from Japan's Metabolist Movement in architecture. Arising in the late 1950s and flourishing until the early 1970s, the movement was a utopia-driven project aimed at realizing the city as an organic process through the development of infrastructure capable of flexible growth and infinite adaptation for an anticipated postindustrial age.[52] That Mori did not turn to the Metabolists for inspiration has to do with the movement's infrastructural vision: despite the movement's name, its vision adhered mostly to an anatomical model that imagined centrally planned megastructural complexes of interchangeable modular components.[53] In other words, the Metabolist movement thought about expansion, but not about the urgent necessity for a system without limits that could accommodate extreme events.

Instead, Mori drew inspiration from life-sciences texts that were also inspiring ALife practitioners. This is perhaps not surprising considering their shared concern for overcoming centralized schemas and realizing emergent complexity. Mori's specific interlocutors were Richard Dawkins, whose 1976 publication *The Selfish Gene* reimagines evolution as a genome-centered, distributed, computational process in which the body of a living organism serves merely as a vehicle of genomic replication, and Gustav Nossal, who redefined the immune system as a highly distributed, complex, adaptive network in which immune cells respond to threats in a localized but globally coordinated manner, with each immune cell adapting and responding on its own without centrally issued directives.[54] In addition, he was influenced by the German biologist Jakob von Uexküll, whose theories of species-specific embodied perception, or *merkwelt*, became important in the evolution of ALife into "New AI."[55] Without moving to claims of subjectivity—thus avoiding phenomenology—Uexküll argued for understanding perception and cognition as interrelated and corporeally distributed phenomena. Largely in debt to a theory of autopoiesis introduced in 1972 by the Chilean biologists and philosophers Humberto Maturana and Francisco Varela, this literature helped redefine life in emergent terms as the property of a distributed, self-organizing system resilient to extreme environmental events.

According to Mori, these thinkers overturned a dominant, reductionist view of living organisms as centralized systems in which each cell has a task-specific function within subsystems that perform under strict directives from the brain. He understood these thinkers as offering instead a schema of distributed autonomy in which there is no master-servant hierarchical relation, but rather "a continuous cooperative interaction or reciprocity between cells" (*saibō wa otagai taitō na kankei de kyōchō shiatte karada zentai o tsukutte iru*). Under this approach, living organisms are reconceptualized as networks of locally focused but globally coordinated operations that allow for the integration of radical contingency into the emergent order. Or, as Mori puts it, distributed autonomy allows for treating "irregularity as regular" (*ijō ga seijō*).

Importantly, drawing from similar sources, ALife researchers reached the same conclusion in their attempts to realize a self-organizing, emergent technological system. Echoing Mori, they began to perceive irregularity as a principal property of self-organizing emergent complexity in nature. In contrast to conventional theories of natural evolution, in which the emergence of irregularity is an engine of change but figures as an anomalous event within otherwise stable systems, ALife worked from the counterintuitive premise that nature is characterized by metastability, not homeostasis, making irregularity necessary for self-organization and complexity. According to ALifer Langton, the optimal point for dynamism within any system (economic, social, and so on) is thus order at the "edge of chaos."[56] While ALife researchers applied these ideas to developing emergent complex virtual entities, and Brooks used them to build autonomous robots, Mori put them to work through distributed autonomy to fabricate operating systems for large technical ensembles capable of emulating organic processes of growth, metabolic renewal, and immunity as technological reliability, maintenance, and expansion. Somewhat ironically, albeit in a way that perhaps prefigures the evolution of ALife into New AI, the first tangible iteration of distributed autonomy in a transit system took the acronym HAL (recalling the infamous psychopathic AI HAL 9000 from the 1968 Arthur C. Clarke and Stanley Kubrick film and novel *2001: A Space Odyssey*).[57] Eventually the name was changed to ADS (Autonomous Decentralized Systems).

The New Capacity of Extreme Capitalism

When Mori set out to solve the capacity crises of the 1970s through "irregularity as regular," he could not have imagined the broad resonance of this innovation. In adapting a metabolic schema of operation that treats irregularity as regular for a technological ensemble, Mori enacted a method of thinking analogically that is fundamental to the ideal of becoming organic. The result is a novel technological system—ATOS—that, while not organic per se, demonstrates a certain level of expanded margin of indeterminacy and a corollary degree of concretization that allow it to be designated neo-organic. For Mori, creating a system capable of accommodating irregularity within its regular operation realized an objective for fault-tolerant technology, which I have parsed as extreme infrastructure.

On the one hand, in Mori's thinking irregularity takes the form of a surge in capacity that puts an overwhelming burden on a technical system. Irregularity in this instance is a disordering force and therefore must be both accommodated and contained. On the other hand, through the analogy with metabolic process, irregularity takes on a potentially positive meaning, referring to a surge in demand resulting from an intensification of and growth in activity. Irregularity, in this regard, is the force behind self-organizing emergence and thus the index of a kind of vital energy and flourishing. As such, it becomes an expression of the protean quality of a healthy metastable system in ways that align it with what Melinda Cooper argues is the formulation of a novel neo-organic configuration of capital and what I contend is the basis of extreme capitalism.

Cooper's argument covers some of the same global crises I have laid out here. However, she focuses on delineating the specificity of post-1970s neoliberalism as distinct from Foucault's exposition of the birth of neoliberalism around capitalism's crises of the 1930s and 1940s. For Cooper, contemporary neoliberalism's particularity derives from its conjunction with the development of nonlinear, complex computer modeling and ALife's counterintuitive formulation that nature is characterized by metastability rather than homeostasis.[58] This reconceptualization of nature, she argues, provided an alternative interpretation, in contrast to the Club of Rome's dismal forecast for the future in the wake of the economic, social, and environmental crises of the late 1960s and early 1970s. Whereas the Club of Rome's forecast was predicated on an equilibrium model of nature, by embracing metastability the reconceptualization of nature developed a neo-organicist logic that transformed crisis from aberration and extreme event into an engine of dynamism and emergent complexity. Consequently, the idea of a *limit* to growth (environmental, economic, or otherwise) was discredited. If nature is a dynamic self-organizing system capable of ceaseless and recombinant emergences, the argument goes, then crises are indexes of vital regeneration, not limit. Capacity, in other words, was transformed from a problem of maximum performance and material thresholds into a measure of emergent complexity. Tracking the transposition of this neo-organicist logic through a series of legislative and institutional decisions in the economic sphere, Cooper finds its articulation in a novel structure of speculative finance imbricated with biotechnology that capitalizes on the dynamic protean quality of organic life. The critical corollary is that, in neoliberalism, the location of value production is shifted from the labor of the living to the dynamism of life itself.

There have been several theories of economic value in the information age, most of which stop at the idea that information is a new and immaterial commodity form, thus resolving the material and environmental restrictions of material commodity production and consumption under industrial capitalism. Michael Hardt and Antonio Negri famously try to go beyond this when they argue that, in the passage from a Fordist to a post-Fordist economy, labor is recast from the surplus labor of laboring bodies in commodity production to immaterial labor that tends to be network-based, building on new modes of communication enabled by the global information network.[59] Cooper asks us to consider the way creation of capital shifts entirely away from labor (material or immaterial) to become located in surplus processes of life cultivated at the intersection of biology and technology. Under the new schema, economic production takes place "at the genetic, microbial and cellular level."[60] What is more, the new schema makes irregularity, in the form of socioeconomic instability, a force to be cultivated in the interest of emergent economic dynamism, not something to be contained. Whereas Cooper follows the production of a neo-organicist logic in biotechnology, I am interested in its infrastructural iteration as extreme capitalism.

Extreme capitalism inverts the premises of extreme infrastructure's capacity. For Mori, the central challenge facing technical systems in the 1970s was how to accommodate the growing economic intensity of an increasingly dynamic and unpredictable world and market. In Mori's formulation, irregularity born of that dynamism and unpredictability needed to be attended to without risking systemic failure. Accordingly, the demand was for technological systems able to keep pace with the emerging 24/7 pace of economic activity. The challenge was about designing a resilient system with a broad performance-capacity spectrum in order to contend with irregularity. But under the neoorganicist logic of extreme capitalism, dynamic unpredictability is no longer a force that infrastructure is expected to contain; rather, infrastructure is asked to *generate* it.

Extreme infrastructure has become the medium of an extreme capitalism that works to produce and capture the protean dynamic of complex collectives as new forms of monetary value. Extreme capitalism articulated through Tokyo's commuter train network is thus not only about increasing the scale of production and consumption; it is also a perversion of machine theory for generating a novel, frictionless synergy of human and machine toward boundless consumption.

Beyond Labor

By the time JR East began adapting Mori's principle of distributed autonomy to create ATOS, the notion of irregularity had taken on additional meaning as a result of the perception of a new challenge for value production in the commuter train network. The director of JR East's Frontier Service Development Laboratory distills the nature of the new challenge in an article for *JR East Technical Review*:

When daily activities of people were relatively stereotypical, demand for transportation was appropriately forecast through investigation of changes in population or economic figures, and it was possible to predict with reasonable accuracy the number of JR East customers. However, when diversification of value perspectives, ways of thinking, and activities become significant, and societal systems and frameworks change as seen in today's Japan, it becomes difficult to predict demand for transportation and other associated activities.[61]

"When daily activities of people were relatively stereotypical" is a gloss for mass-mediated postwar Japan. The director's argument is that the commuter train network can no longer rely on a steady revenue as a vehicle of mass transportation in the same way it did during the postwar era. It must instead work to create new sources of value under circumstances in which irregularity has become regular. Irregularity, in this regard, refers not to malfunction or surges in mass demand but rather to emergent and diverse activities and ways of thinking among the urban population. Economic value, in other words, is no longer a surplus extracted in the conveyance of bodies to institutions of labor, but rather a derivative of diverse (read: irregular) lifestyle patterns. Mori also reiterated this argument when we spoke, contending that, under the new socioeconomic conditions of post-1970s information-society Japan, JR East needs to transform itself from a transportation company into a "service-creation company" that creates value by "enriching our lives" (*ware ware no seikatsu o yutaka ni suru*).

The director goes on in the article to propose systems for new forms of value creation that will transform the commuter train network from a means of transport to and from places of labor, education, and consumption into a destination in itself. What she envisions is essentially an environment that has been realized in recent years through the renovation of JR East train stations throughout Tokyo. In contrast to the train terminal/department store model that dominated the growth of urban railroads throughout the twentieth century, under the new paradigm the area within the ticket gates, which was a space designed for optimum flow in accordance with the principles of mass transportation, is transformed into a consumer gallery packed with boutique shops and specialty food stores. The resulting atmosphere is something between a shopping mall and a theme park, complete with performances, shows, and a train ride (instead of a roller coaster). The recent reconstruction of Mitaka Station and Tachikawa Station on Tokyo's Chūō Line exemplifies this transformation. In both cases, between the ticket gates and the platforms, commuters must pass a wide variety of bakeries, bookstores, cafés, clothing stores, delicatessens, and gift stores. Mori, it is worth adding, views these changes as positive innovations. But he wants them to go even further. He envisions a future in which the train station will become a city within the city by consolidating within its confines every imaginable facet of daily-life service (many of which are already part of the train-station complex), from fitness gyms to nursery schools to hospitals to city administration. That this model contradicts his emphasis on the benefits of decentralization did not seem to bother him.

Toward Boundless Synergy

Turning train stations into festive shopping zones was only the first step in the transformation of Tokyo's commuter train network and was ultimately insufficient for creating the kind of new sources of value that the director of JR East's Frontier Service Development Laboratory deemed necessary for the company to survive. The company needed to transform its network into a vehicle of extreme capitalism, capable of producing and transforming capricious consumer desire into economic value. It found what it needed with the "Super Urban Intelligent Card," otherwise known as SUICA. Developed by one of Mori's former students, SUICA deploys the principle of distributed autonomy used in ATOS for the creation of a digital ticketing and payment system and is part of an emergent ecology of symbiotic distributed-autonomy systems being engineered for the city.

SUICA replaces the conventional paper train ticket and the commuter pass with an integrated circuit (IC) chip embedded in a SUICA card, credit card, smartphone, or *keitai*. SUICA also serves as an electronic wallet usable for purchases inside the system and, increasingly, throughout the city. When the integrated IC chip comes into proximity with the radio frequency emitted from a sensor, bidirectional communication at 250 kilobytes per second is initiated with a distributed-processing computer system that reads and authenticates the card, calculates the fare, and rewrites the subtracted amount, all of which takes place in less than 0.2 seconds. The process allows commuters to enter the train system or make a purchase by simply tapping a wallet or purse on a sensor at the ticket gate or at a store. A basic fee is deducted when one enters the system, and the full fare is deducted when one exits.

With SUICA, "irregularity" comes to mean consumer impulse. There is no need to plan a destination or a purpose for travel before entering the system. As long as one's SUICA is linked to a bank account or credit card, the system promises frictionless mobility and spontaneity to fuel impulse consumption. It is about being unhindered by the restrictions of space and time that shaped the flows of people and commodities in cities throughout the twentieth century. SUICA produces and directs contingency toward emergent economic synergy. Since no passage through the system need be determined at the outset, every journey is seemingly open to endless possibilities, to sudden whims to change directions, to explore, or to pursue a chance encounter. When all train and metro lines in the Kanto and Kansai areas adopted compatible systems in 2010, all spatial delineations that had been produced within the city via the transportation network vanished into a single seamless flow of transitory consumption.

"I Live with SUICA"

Nothing demonstrates the drive to produce unbounded consumer flows and desire through SUICA better than a series of television commercials for the technology that aired around its debut in 2001. The commercials follow the day-to-day adventures of an attractive, fashionable woman in her mid-twenties and her cute penguin companion. The penguin is the official SUICA mascot, chosen for the onomatopoeia used to express its smooth movement through water—*sui sui*—as a way of conveying the promise of frictionless mobility through urban system space with SUICA. Fulfilling the role of something between a lover, child, and partner, the penguin is presumably the woman's sole companion. Yet, for the most part, the woman and penguin do not seem lonely or alienated from the social spheres of their respective species. The two live together in a traditional one-story wooden home with a small garden in the back, which the penguin tends to in the morning while the woman sleeps. We see them in the commercials enjoying a seemingly carefree life of fun and adventure in and around Tokyo's commuter train network, to the accompaniment of cheerful background music. They ride the train together, take walks in quaint neighborhoods, shop for quirky items, sightsee in rural locales, and so on. Each outing involves some kind of impulse purchase, such as buying the penguin's favorite treat (fish sausage, of course), going through a commuter-train-network ticket gate, or buying clothing. In each instance, the young woman handles the transaction effortlessly, with a touch of her SUICA card at a sensor. The penguin holds on to the SUICA card or Mobile SUICA smartphone for the woman and hands it to her for the critical moment in each commercial when the technology is used for a purchase or to access the commuter network. The commercials always end following this moment with the woman declaring "I live with SUICA" (*watashi wa SUICA to kurashite imasu*) as the phrase appears in small, feminine, handwritten characters over a scene with a passing train. The phrase carries plain double meaning, referring to the facts that the woman actually lives with SUICA the penguin mascot and the Super Urban Intelligent Card provides the means for her to pursue boundless consumption in her carefree, self-satisfying lifestyle.

Entirely absent from the SUICA commercials are any references to the continuing reality of commuting. There are no scenes of packed trains or crowded platforms. The commercials promote instead the image of the commuter train network as a medium for the realization of one's customized lifestyle through frictionless mobility, spontaneity, and impulsive consumption. The emphasis in this regard is on leisure, play, and adventure as modes of consumption that are all available within the space-time of the commuter network.

Should the image of frictionless space through SUICA begin to appear attractive, we need only remember Gilles Deleuze's caveat that in societies of control, "what counts is not the barrier but the computer that tracks each person's position—licit or illicit—and effects a universal modulation."[62] Indeed, SUICA's guarantee of spontaneity and emergence is about the massive amount of data produced in and by the commuter traveling through the commuter train network assemblage. Culled from ticket gates, the data is integrated into *daiya* production for ATOS. More importantly, it is analyzed in order to further produce emergent lifestyle trends that can translate into new sources of value extraction.[63] Intended to allow JR East's marketing group the ability to "understand and respond to the needs of each commuter on a deep level" (*okyaku sama no nīzu [needs] ni rikai o fukame*), the data allows for a constant calibration of commodities, services, and advertising at every point throughout the system.[64] Mass advertising is thus superseded by pinpoint advertising strategies organized around commuter profiles whose consumer capacities, desires, and whims have been rendered part of the big data produced with SUICA. Although JR East has held back on mobilizing the system's full power over concerns of personal-data leakage, in the near future it will be no mere coincidence when a commuter on the Chūō Line encounters an advertisement for a new car the day that he or she has finished paying a home loan and has cash to spare.[65] JR East's marketing division will know which train the specific commuter rides, the kinds of snacks and drinks he or she is in the habit of purchasing, and whatever reading material he or she buys from any one of the kiosks or vending machines on the platform.

Insofar as JR East has been unable to operationalize the full emergent value of SUICA, the system's inherent potential speaks to an underlying impulse at play in JR East's reinvention of its network, namely to transform the commuter train system into an embodied network space that behaves according to the look, feel, and experience of the virtual networked space of the internet. Accordingly, the objective is to produce, by virtue of subtle tracking mechanisms embedded in the architecture, a commuter experience with the same sensation of coherence that is created when one moves across heterogeneous websites. Just as searching the web for a new bicycle, for example, triggers mechanisms that cause advertisements for bikes to appear in the sidebar of every subsequent page visited, making a purchase in the commuter train network will have the same effect, producing the uncanny sense of something sentient within the apparatus, something able to anticipate one's still-inchoate desires.

A similar impulse for anticipatory marketing drives projects throughout the world to build so-called "smart" infrastructure. As with ATOS, the formative perception for these projects is the idea that, with the development of information-network technology, the global economy has transformed into a complex, evolving system that produces an unprecedented degree of indeterminacy in the form of emergent patterns of behavior. Indeterminacy, in this context, can mean either diversification of consumer demand or environmental instability. Both qualities of indeterminacy place similar requirements on the system, demanding that it be capable of constant operation despite the malfunction of components and capable of infinite capacity for rapid and adaptable growth in correspondence with the tempo of economic change. Although the term "smart" suggests something like artificial intelligence and an ability to engage users on a cognitive level, what is "smart" about this technology is more commensurate with notions of ALife. It has to do with the capacity to generate life as emergent patterns of embodied behavior, not as sentience and intelligence. So-called smart systems do not think. They generate synergy between humans and machines and then capture that synergy as economic value. They do so, in part, by sensing fluctuations in the environment and responding to gestures, motions, and the congestion of bodies and things. The "intelligence" of such systems thus operates on a more affective level. It is attuned to shifting intensities within the urban ambiance as it works to realize the city as a form of emergent, living organism, in which the symbiotic interaction between distributed systems enables the city to sense and respond as a global entity to emergent patterns within its internal milieu. Such systems are what Nigel Thrift is talking about when he argues for recognizing the emergence of new materiality in which constant background computing embedded in our everyday architecture exists as a kind of "technological unconscious," creating the experience of architectural space as responsive and fluid, bending to user needs.[66]

The Post-Organic

This chapter ends as it began: with the questions "What went wrong?" and "How has the organic become complicit with capitalism's compulsion for extreme consumption?" As I suggested at the beginning of the chapter, the conceptual integrity of the term *technicity*, as a way in which to think about a techno-ethics that moves beyond the overly general term *technology*, is at stake here. For if the capacity to realize a greater degree of organic-esque concrescence—a "becoming organic"— lends itself to resilient infrastructure at the same time that it facilitates even greater levels of consumption, then the term would seem to no longer provide the kind of critical intervention for thinking toward a different relationship with machines. For Simondon, changing the structure of labor under capitalism was one and the same with changing our relationship with machines by recognizing the hylomorphic model's problems.

I am not ready to give up on the term *technicity*, however, just as I am not ready to give up on the term *emergence*. Both are powerful concepts that enable us to not only think about but also demand a world more suited to collective individuation. In an attempt to hold onto these concepts, it is worth considering the possibility of an implicit flaw with my question, "What went wrong?" The question assumes, to some extent, that technicity and emergence can no longer serve as conceptual resources for a subversive countermovement to capitalism's seemingly endless capacity for producing desire and consumer subjectivities to go with that desire. This is indeed what Thacker and Galloway imply when they declare, "We're tired of rhizomes."[67] The problem is that, metaphorically speaking, rhizomes have become much like the trees, whose rigid, molecular, and hierarchical structure was supposed to provide a position from which an ever-adaptive, modulating, and dynamic rhizomatic topology could mount a subversive resistance. The problem of the rhizome becoming politically and culturally commensurate with an arboreal modality is analogous to the dilemma faced by a revolutionary force that liberates the land from the rich and redistributes it to the landless only to discover soon enough that yesterday's revolutionary landless have become today's conservative landowners.

The solution that followed in places like China was to pursue a per-petual revolution that demanded the constant relabeling of former revolutionary elements as anti-revolutionary and the creation of new revolutionary elements to displace them. We risk falling prey to the same compulsive cycle if we insist that what technicity and emergence offer are subversion. To continue with the analogy, if we insist on sus-taining subversive resistance, we must quickly develop the next set of post-emergence, post-technicity terms while holding academic court to condemn Simondon and Deleuze along with numerous contemporary STS scholars for their misguided (albeit unwitting) complicity with cap-italism. Certainly, there must be an alternative approach.

I find it helpful to recall here Michael Goddard's discussion of Félix Guattari's media ecology concept. In Guattari's work, ecology is a ca-pacious term that includes all manner of systems (social, environmen-tal, mental, and so on) that operate as subjectivity-producing milieus.[68] In this context, suggests Goddard, we can read Guattari's interest in the potential media ecologies generated through such technologies as radio as less about "the subversive use of a technical media form than the generation of a media or rather post-media assemblage, that is a self-referential network for an unforeseen processual and politi-cal production of subjectivity amplifying itself via technical means."[69] In other words, the point of a media ecology does not specifically have to be resistance. Rather, it just has to provide a dynamic milieu of novel material and immaterial arrangements that lend themselves to furthering critically driven political and cultural individuations. The same can be said of technicity and emergence. That technical ensembles demonstrating closer alignment with what Simondon calls "becoming organic" lend themselves to an increasing capacity for frictionless syn-ergies of capital does not necessarily mean we must abandon technicity and the value it places on technological concrescence. It does demand, however, that we remain wary of this capacity and use it as an impetus to constantly question the collective limits and possibilities of milieus we engender in conjunction with technological innovation.

Notes

Some of the material in this chapter has also appeared in Michael Fisch, "Tokyo's Commuter Train Suicides and the Society of Emergence," *Cultural Anthropology* 28, no. 2 (2013): 320–43; and Michael Fisch, "Remediating Infrastructure: Tokyo's Commuter Train Network and the New Auton-omy," in *Infrastructure and Social Complexity: A Routledge Companion*, edited by Penny Harvey, Casper B. Jensen, and Atsuro Morita (London: Routledge, 2016), 115–27.

1. Kitahara, Kera, and Bekki, "Autonomous Decentralized Traffic Management System."
2. Fisch, "Tokyo's Commuter Train Suicides and the Society of Emergence."
3. For an excellent analysis of current online customized advertising, see Martinez, *Chaos Monkeys*, 382. Also, for an in-depth exploration of the logistics of just-in-time global commodity production and delivery, see LeCavalier, *The Rule of Logistics*.
4. My thinking here is informed by Thomas LaMarre's discussion of technical optimization in the creation of anime; see LaMarre, *The Anime Machine*, 140–41.
5. Haraway, *When Species Meet*.
6. With the term *becoming-organic*, I borrow John Johnston's phrasing of Simondon's concept. Johnston, *The Allure of Machinic Life*, 7. For Simondon's exposition of technological progress as becoming-organic, see Simondon, *On the Mode of Existence of Technical Objects*, 71–81.
7. "Technical Mentality," in Simondon, *Gilbert Simondon: Being and Technology*.
8. Cooper, *Life as Surplus*.
9. Halpern, *Beautiful Data*.
10. Schivelbusch, *The Railway Journey*.
11. Isamu, "Ressha shūchū seigyo sōchi (CTC) no kaihatsu."
12. JR East's Kanto train network includes twenty-four train lines, eighteen of which were under ATOS control from the *shireishitsu* in 2005. By January 2012, twenty of those train lines had been fitted with ATOS while plans were underway to expand ATOS on the remaining four train lines. See Ito, "Development and Update of ATOS," 33.
13. JR East employee, conversation with author during a tour of the *shireishitsu* in November 2005.
14. I received permission to tour the facility through Tomii Norio, to whom I was introduced by employees at the RTRI Cerajet Division, where I was teaching English. The permission I received was contingent on my promise not to take photographs or publish any sensitive details regarding the facility.
15. Ito and Hideo, "Autonomous Decentralized System with Self-organizing Function."
16. Foucault, *Discipline and Punish*, 205.
17. Foucault, *Security, Territory, Population*.
18. Galloway and Thacker, *The Exploit*, 153.
19. There are any number of places one can find examples of this discourse, but some of the most explicit articulations appear in celebratory treatises on information society, such as William J. Mitchell's *Me++: The Cyborg Self and the Networked City* (Cambridge, MA: MIT Press, 2003).

20. Papers presented at the International Workshop on Autonomous Decentralized Systems or at the International Symposium on Autonomous Decentralized Systems discussed and proposed diverse applications of the technology, from factories to monetary management systems to railroad traffic operations. These papers were delivered and published in English.
21. See for example Kitahara, Kera, and Bekki, "Autonomous Decentralized Traffic Management System."
22. Mito, *Teikoku hassha*, 116.
23. Yamamoto, "Sekai ni hirogaru jiritsu bunsan."
24. In reality, customers tend to place orders at *kaitenzushi* restaurants as well, asking the sushi chef for a specific fish.
25. "Infrastructure," *GlobalChange.gov*, accessed June 18, 2016, http://nca2014 .globalchange.gov/highlights/report-findings/infrastructure.
26. Graham and Marvin, *Splintering Urbanism*.
27. Some economists perceived the crises to be the result of the "built-in limitations" of the postwar Keynesian model of political economy and industrial growth. See Castells, *The Rise of the Network Society*.
28. Meadows et al., *The Limits to Growth*. Melinda Cooper, whose work I discuss later in the chapter, offers an excellent analysis of this document.
29. Morris-Suzuki cites a member of the Japan Computer Usage Development Institute—a think tank that pressed in the early 1970s for a transition to an information economy—as referencing the report in a claim for information's value as a nonpolluting commodity. Morris-Suzuki, *Beyond Computopia*, 60.
30. Ibid.
31. George, *Minamata*.
32. For excellent analyses of these demonstrations and movements, see Sand, *Tokyo Vernacular*; Sasaki-Uemura, "Competing Publics."
33. Morris-Suzuki, *Beyond Computopia*, 57.
34. I follow Morris-Suzuki here in using the term *pro-monetarist* instead of *neoliberal*; see Morris-Suzuki, *A History of Japanese Economic Thought*. The former term more accurately conveys the particular neoconservative orientation that inflected economic thinkers in Japan who were heavily influenced by Milton Friedman's neoliberal economic theories.
35. Weathers, "Reconstruction of Labor-Management Relations in Japan's National Railways," 624.
36. NTT was privatized in 1985, two years before the final privatization of JNR. The battle throughout the 1970s and 1980s over the reorganization of JNR had indeed set the stage for the privatization of NTT. The process of privatizing the Japan Postal Service Agency began with the formation of Japan Post in 2003. In 2007, Japan Post was privatized to form the Japan Post Group.
37. My summary here of the attempts to institute changes in JNR draws from a number of sources, including Weathers, "Reconstruction of

Labor-Management Relations in Japan's National Railways"; Ishikawa and Imashiro, *The Privatisation of Japanese National Railways*; Kasai, *Japanese National Railways*; Watanabe, "Restructuring of Japanese National Railways."

38. Weathers, "Reconstruction of Labor-Management Relations in Japan's National Railways," 624.

39. I am in debt to Hirakatsu Tatsuo, Kawabata Kazuo, and Sakai Naoaki, former JNR employees who participated in a twenty-year protest, for taking the time in July 2008 to explain the fate of Kokurō members following privatization to me.

40. David Harvey offers a brief history of the financial situation at the time. See Harvey, *A Brief History of Neoliberalism*.

41. Takeuchi, "*Seimei ni manabu*"; Yamamoto, "Sekai ni hirogaru jiritsu bunsan."

42. NHK's *Project X* documentary on the invention of the first electronic ticket gates (chapter 1) states that the maximum was eighty-eight commuters per minute.

43. Gleick, *The Information*.

44. Gleick, *Chaos*, 216.

45. Ross Ashby, a central figure in postwar cybernetics, developed the first mathematical proof for self-organizing systems in 1947, in "Principles of the Self-Organizing Dynamic System." See Ashby, "Principles of the Self-Organizing Dynamic System." According to John Johnston, we can see Ashby's concept later taken up by computer scientists trying to develop Artificial Life.

46. Ziman, *Technological Innovation as an Evolutionary Process*, as cited in Johnston, *The Allure of Machinic Life*.

47. Parikka, *Insect Media*.

48. Langton, ed. *Artificial Life*.

49. Brooks, "Intelligence without Representation."

50. Ibid., 157.

51. Brooks, "Intelligence without Representation," 145.

52. Lin, *Kenzo Tange and the Metabolist Movement*.

53. For an account of the cybernetic influence on the Metabolist movement, see Wigley, "Network Fever."

54. Dawkins, *The Selfish Gene*; G. J. V. Nossal, *Antibodies and Immunity*.

55. See Johnston's discussion of Uexküll in relation to what he defines as "the new AI." Johnston, *The Allure of Machinic Life*, 347.

56. Ibid.; Cooper, *Life as Surplus*. Note that while Cooper attributes the phrase "edge of chaos" to Stuart Kauffman, Johnston associates it with Christopher Langton. As Johnston commits to an exhaustive analysis of the ALife project, I have chosen to follow his reading and relate the term to Langton.

57. Miyamoto et al., "Autonomous Decentralized Control and Its Applica-

tion to the Rapid Transit System." The system was developed for twelve kilometers of the Kobe City subway system. HAL stands for "Harmonious, Autonomous, and Localities."

58. Cooper, *Life as Surplus*, 43–44.

59. Hardt and Negri, *Multitude*.

60. Cooper, *Life as Surplus*, 19.

61. Egami, "Idō to seikatsu ni okeru aratana kachi no kōzō wo mezashite— kachi · kaiteki · kukan no kōzō."

62. Deleuze, "Postscript on the Societies of Control."

63. A JR East report puts the number of SUICA cards in use at 42.47 million as of March 31, 2013, up from 38.88 million cards as of March 31, 2012. See JR East, "Review of Operations: Non-Transportation," 2013 Annual Report, www.jreast.co.jp/e/investor/ar/2013/pdf/ar_2013_10.pdf; and JR East, "Review of Operations: Non-Transportation—SUICA," 2012 Annual Report, www.jreast.co.jp/e/investor/ar/2012/pdf/ar_2012_10.pdf.

64. Egami, "Idō to seikatsu ni okeru aratana kachi no kōzō wo mezashite— kachi•kaiteki•kukan no kōzō."

65. According to JEKI project managers, the marketing group decided not to collect personal data for fear of commuter backlash over concerns of invasion of privacy. Similarly, in response to voiced concern from commuters, JR East was also forced to allow SUICA users to opt out of data tracking by registering their card number online. Still, one needs to opt out rather than opt in. Moreover, the availability of such an option is not widely known; it is only available via careful exploration of the JR East website. (Yoshihara Mihoko, interview by the author, June 16, 2005, at the JEKI office in Ebisu, Tokyo.)

66. Thrift, "Movement-space."

67. Galloway and Thacker, *The Exploit*, 153.

68. Guattari, *The Three Ecologies*.

69. Goddard, "Towards an Archaeology of Media Ecologies," 14–15.

FOUR

Gaming the Interval

In the single packed-train scene of Ichikawa Kon's 1957
film *Man-in densha* (The full-up train), an advertisement for
the film hangs at center screen, just above the heads of the
cramped commuters (figure 4.1). Alluding to the indivisi-
ble relationship between mass transportation, mass media,
and conditions of mass production, the advertisement sa-
tirically points to the film's involvement in the conditions
of mass-mediated capitalist society it sets out to critique.
Similarly, the advertisement gestures toward a representational correspondence between the railroad and cinema as
foundational and co-articulating vision machines of the
twentieth century. Immobilizing bodies while mobilizing
vision, cinema and trains are both understood to have produced commensurate experiences of immersive technological mediation in which the participants—moviegoers
and railroad passengers respectively—emerge as analogous instantiations of modern subjectivity.[1] That cinema
became the quintessential medium through which the
shock, temporality, and tedium of the commuting experience becomes legible is often attributed to these structural
and phenomenological affinities.

The cinematic trailer for the 2005 international-hit film
Densha otoko (*Train Man*), directed by Murakami Shōsuke,
brings into focus a different techno-configuration by introducing the web into commuter space. Emerging from
a discussion thread on the Japanese-language subculture
2channel textboard website, *Densha otoko* tells the supposedly true story of a virtual community that coalesces to

FIGURE 4.1. Inside the train in *Man-in densha.*
Source: *Man-in densha.* Directed by Ichikawa Kon. Tokyo: Daiei Studios, 1957.

facilitate a romance between one of its otaku[2] participants and a woman that was sparked by an incident on a Tokyo commuter train. Plucked from the web by Japan's publishing-media powerhouse Shinchōsha, *Densha otoko* became a national sensation in 2004 from the moment of its book debut, selling 260,000 copies in just three weeks and 500,000 copies in two months. Japan's media-entertainment industry moved quickly to capitalize on the story with manga and television-series versions, while Tōhō Cinema produced a feature-length film of the story that became an international hit. Commuter train and web converge in *Densha otoko* to offset the relationship between train and cinema that is articulated in *Man-in densha.* More importantly, *Densha otoko* presents the web as eliciting a novel articulation of collective through its remediation of the space and time of transit. This convergence and remediation is allegorically invoked in the film's trailer, where we see the otaku and the woman sharing a train car as texts from the website's discussion thread stream by the window. Industrial capitalism's foundational network of transportation is subsumed, the images suggest, by postindustrialism's information network, rendering the train a stage for a postmodern play.

In 2004, when the book and film of *Densha otoko* debuted, that kind of fluid symbiosis between the train network and the web was more fantasy than reality. Full web access via the smartphone was still a few years in the future, and the mobile wireless-communication market was entirely dominated by a closed *keitai* system of third-generation (3G) devices. Insofar as these devices offered commuters personal communication through *keitai* mail messaging, they provided access only to a pared-down *keitai*-specific network within the web called the "mobile web." Nevertheless, for commuters who had grown up with the silence of the packed train and commuting as an experience of inevitable transitory disconnection, even this limited connectivity seemed to open the space of the transportation network to the emerging techno-economies and anticipatory social discourses of the information network. *Densha otoko* thus simultaneously spoke to the anticipatory desires and concerns of an urban population that was growing accustomed to getting online and asked them to reimagine the space and time of the commuter train network with the web.

It is this reimagining that concerns me in this chapter. To put it differently, this chapter asks how the web thinks the margin of indeterminacy of the commuter train network. This is not a question of how the web transforms commuting practices. Rather, to ask how the

FIGURE 4.2. Inside the train in *Densha otoko*.
Source: *Densha otoko*. Directed by Murakami Shōsuke. Tokyo: Toho Production Co., Ltd, 2005.

web thinks the train is to consider the ways thinking the train with the web allows for imagining the space and time of commuting differently. That imagining, I argue, transpires as a remediation in both the conventional and expanded (*re-mediation*) sense of the term. On the one hand, it is an attempt to remedy the solitudes, perspectives, and social mores that have become part of the commuter-train space. On the other hand, it is quite literally, a re-mediation: one form of mediation is subjected to the register, logic, and perspectival associations of another system of technological mediation.

In asking how the web thinks the train, I explore three different instantiations in which the commuter train is remediated via the web. The first is a web-based hypertext novel titled *99 Persons' Last Train* (*Kyūjūkyū nin no saishū densha*) that attempts to simulate a different experience of commuter space through hypertext structure. The second is the story of *Densha otoko*, which I read as an attempt to remediate, in virtual space, a missed collective encounter on the train. Finally, the third is a *keitai* game titled *Days of Love and Labor* (*Ai to rōdō no hibi*), which develops a critique of capitalism by encouraging a kind of thinking with the train. I order these examples by proximity established between the web and the space-time of the train, from least to most. Although the hypertext novel *99 Persons' Last Train* adopts the train system for its setting, it is meant to be read on a computer through a web browser and is thus spatially and temporally removed from its subject. By contrast, the story of *Densha otoko* begins with an actual event on a commuter train that impels the emergence of a virtual collective. Finally, *Days of Love and Labor* puts the player within the immersive, mediated milieu of the train car. In sum, as the chapter advances, we think the train with the web, moving progressively inward, from outside the train into the train and the actual space and time of commuting.

In each example I explore, the remediation of the train via the web takes its cue from web-based computer gaming. Through the computer game, the commuter train is transformed from a representational vehicle for satirical sociopolitical critique and modern subjectivity into an interactive and dynamic platform for simulation. As the game creator and theorist Gonzalo Frasca writes, simulation and representation are two different but overlapping ways of dealing with reality that each offer different rhetorical possibilities. Whereas representation is about "depicting, explaining and understanding reality," simulation produces an experience of a dynamic system in accord with the users' interactions.[3] In other words, simulation foregrounds the ontological open-ended character of material situations. In a driving simulator, this would involve introducing a driver to the basic mechanical systems of the car and the challenge of interacting with the often unpredictable flows of traffic. Even if the unpredictability of the simulation is limited to certain programmed parameters, for the simulation user it feels like a personal experience over which he or she has a certain amount of control. The effect is to generate and immerse a user in a form of alternative reality with the potential for different outcomes.[4] The aim of this effect is often, but not always, pedagogical. Driving simulators are supposed to teach you how to drive by giving you the opportunity to respond physically and not just mentally to various road conditions. A computer-game simulation may not be instructional at all; it may just be about the pleasure of visual-cognitive stimulation. When a simulation is pedagogical, however, it encompasses a performative dimension. Performativity speaks of an unpredictable and experimental quality. It emphasizes process as a dynamic and interactive phenomenon from which will emerge not only new understandings but also new ways of knowing through becoming, which is to say novel ontogenetic possibilities.[5] The performativity of simulation is thus about a space bounded by certain overarching protocols, yet whose internal processes do not have a determined outcome. The objective of performative simulation is not to win but to transform, which raises the question of what kind of transformation. For the gamer and media theorist McKenzie Wark, game space provides a point of intervention into habituated processes of everyday life under capitalism. In Wark's work, the computer game is an invitation to engage in critical analysis of capitalism, and in so doing transform relations in the world. If web connectivity promises to transform the commute into a kind of game space, I argue in this chapter, then it opens the possibility for the train to serve as a point of critical intervention and transformation that begins with the question, Can the train teach us to care?

99 Persons' Last Train

When the French anthropologist Marc Augé, in his ethnography of the Paris Métro, depicts commuters as "solitude[s]," he captures the underlying paradox of the subway as a collective phenomenon that is experienced differently by each commuter in a correspondingly similar way.[6] Solitude, as such, denotes a shared yet discrete and individual experience. Augé's work on the Paris Métro is in many ways an attempt to develop an "ethnology of solitude," which, as he suggests in the final line of his book *Non-Places* (written almost a decade earlier), may be the only method of analysis adequate to understanding life within the complex zones of technologically mediated relations—such as airports, train stations, highways, and so on—that are increasingly part of social existence.[7] Such zones, Augé argues, are formed "in relation to certain ends (transport, transit, commerce, leisure)."[8] They constitute "non-places" in that they elude classical anthropological notions of space and collective as geographically coherent and historically situated relations that give rise to distinct expressions of cultural identity. Non-places, by contrast, resist conventional representational methods, producing fluid and indeterminate collectives of intensely networked yet highly atomized solitudes. For Augé, being able to grasp the solitudes is the precondition for thinking about the possibility of collective within non-places of transit. But Augé ultimately strains to modify a foundational ethnographic approach into an analytical framework adequate to the task of engaging the collective of solitudes that inhabit the techno-social nonspace of the subway. In so doing, he abandons a claim to empirical ethnography in favor of a subjective and fictional style to elaborate the metro experience.

In his description of the multiplicity of narratives generated within the shared experience of a Tokyo subway car, the Japanese science-fiction, detective, and mystery writer Inoue Yumehito echoes Augé's concern for conveying the lived reality of complex networks. Inoue writes, "Passengers share the same train car and thus the same moment in time and space and yet their language, nuance, psychology and expectations are different such that if we were to read from each one of their perspectives we would end up with an entirely different story."[9] Whereas Augé produces an ethnography that veers toward fiction, Inoue fuses fiction with the dynamics of computer gaming in his hypertext novel *99 Persons' Last Train* in order to simulate the experience of the Tokyo metro. Set in Tokyo's Ginza Line, the work tracks the stories of each of the ninety-nine commuters on two separate last trains moving between the line's terminus stations of Shibuya and Asakusa.[10] The story begins at 11:56 p.m., just as the last trains prepare to depart from their respective stations, and it ends at 12:13 a.m., when the trains meet across the platform at Ginza Station near the center of the line.

Inoue began working on the hypertext novel in 1996, uploading sections as he wrote. The final section was uploaded a decade after he began, in October 2006. The complete hypertext novel remains accessible for free on the web, even though the publishing company that supported the project, Shinchōsha, offers the work in DVD format. *99 Persons' Last Train* was an experiment in narrative form that achieved only limited success for Inoue as well as Shinchōsha. Inoue has not attempted anything similar since completing it. As he explains in an interview linked to the top page of the novel, he intended to produce a new kind of fiction using the dynamic format of the computer game to incorporate the reader's subjectivity into the reading process. As in a computer game, he wanted the reader's experience with the text to follow certain rules, but for the story to be contingent on the particular path the reader selected in navigating the narrative. His initial attempt at creating such a work used conventional print to produce what Inoue calls a "game book" in which readers were prompted to choose whether to go right or left at the end of each paragraph, each choice leading to a different section or page. The project failed, according to Inoue, due to both the spatiotemporal limitations of print media and the weak content. It was too difficult to navigate multiple narratives, he explains, and the plot became too much like a game and less captivating as a story. That Inoue eventually found a fit between subject, structure, and hypertext medium in the subway suggests something of an affinity between the two systems. One recalls, in this regard, Lisa Gitelman's suggestion that the prehistory of hypertext is perhaps found in "the integrated structure and semiotics of Grand Central station . . . (1913), with its routes and signals for trains, its routes and signals for passengers, and the tiny spiral staircase that connects an information booth on one level (suburban transit) with an information booth on the other (interurban transit)."[11]

As in a computer game, the interactive structure of *99 Persons' Last Train* lends the reader the feeling of producing the world and its stories with each click. But since the emphasis is on exploring a simulated world rather than completing the typical game objectives of collecting points, mastering skills, and overcoming obstacles, *99 Persons' Last Train* ultimately feels more like the virtual-world program *Second Life* than a computer game.[12] As with most virtual worlds, the reader/player can navigate *99 Persons' Last Train* through an interactive schema, shown in figure 4.3. On the story's top page, a diagram of the subway line, labeled "Index," appears; six of the eighteen stations on the train line are active hyperlinks. Underneath the "Index" is a rough outline of a single train car, with small red and blue square icons inside the train and on the platform. A legend to the side of the train car explains that red icons indicate women while blue icons indicate men. Elsewhere in the story, we learn that light-green icons indicate children. Above the legend are two boxes, the top one showing the station and the bottom one the diegetic time.

To move in time, one clicks on the time display. Each click advances the clock one minute, and the trains move toward the meeting point at Ginza Station on the "Index." With each advance, more stations become active links for the reader/player to pursue. At each station, new characters join those waiting on the platforms and characters board or alight when the train arrives. To move spatially between stations, one clicks the arrow on the box above the time display or chooses a station from the index. In addition, the reader can click on the "Character Index" in the top menu bar of the web page and choose a character from among the ninety-nine character icons with names in Kanji and

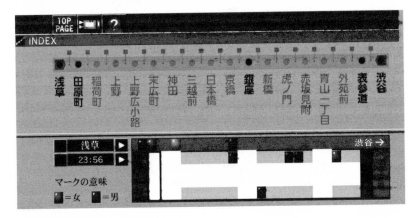

FIGURE 4.3. Schematic of the train car.
Source: *99 Persons' Last Train.* Shinchōsha. http://www.shinchosha.co.jp/99/

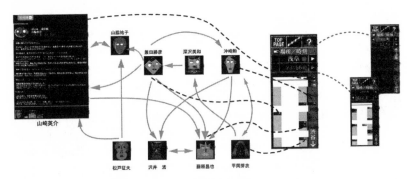

FIGURE 4.4. Schematic of potential perspectives and routes through story.
Source: *99 Persons' Last Train.* Shinchōsha. http://www.shinchosha.co.jp/99/

Hiragana displayed in a menu on the left side of the page. Clicking on a character icon within any of the trajectories opens a page with the character's story, typically composed as a first-person narrative. When one character's dialogue, thoughts, or observations refer to another character, the reference appears as an active hyperlink that the reader/player can pursue if desired, thus shifting instantly to the referenced character's story and point of view. At the bottom and top of each page, links to the selected character's thoughts in the preceding and subsequent minutes appear, and at the end of the text, icons of referenced characters provide hyperlinks to their stories. To help the reader navigate this somewhat complex story matrix, Inoue provides a schematic of potential perspectives and routes (figure 4.4).

In sum, the reader can move through the subway in multiple ways: between characters, within a character's memories, spatially from station to station, or temporally within the seventeen minutes of the story. Each mode of movement carries different rhetorical possibilities. On the one hand, to move from character to character works to reinforce a sense of similarity among the various commuters, or "solitudes," as link after link reveals the deeply unhappy lives of individuals who are all unhappy in similar ways. All are immersed in bitter contemplation, mulling over the events of the day, the caustic remarks suffered from coworkers, or the feeling of having endured general injustices in life. Even raucous salarymen returning home from a night of drinking come off as deeply unhappy, merely dissembling camaraderie with their drunken colleagues. On the other hand, when following a single character's thoughts, one moves inward in a linear fashion toward an increasingly dark and guarded interiority, exposing a process of reflection at the threshold of the unconscious. Such linear movement reconstitutes a classic depth model of human consciousness and a sense of individual uniqueness while lending the reader/player the feeling of listening in on a psychoanalysis session. There is no clearly defined objective to pursuing the links and no sense of movement toward a cure or resolution of conflict for any character. The story progresses rather as a kind of meandering through melancholic expressions of urban life and leaves the reader/player feeling somewhat hopeless.

Moving within *99 Persons' Last Train* according to the spatial and temporal dimensions of the story produces yet another kind of rhetorical register. To progress from station to station along the train line is to move in a linear fashion in accordance with the underlying instrumental telos of the subway as a technological apparatus of modern rationality. At the same time, the hypertext links allow the reader to also move laterally in a nonlinear manner throughout the spatiotemporal dimensions of the story. The latter works to disrupt any underlying structural telos in a way that recalls what Thomas LaMarre calls "relative movement."[13]

The notion of relative movement is the theoretical core of LaMarre's analysis of Japanese animation. It refers to a feeling of movement that is subjectively induced in relation to one's position.[14] For LaMarre, the specificity of Japanese anime is precisely its heavy dependence on lateral movement over movement into depth, or what LaMarre calls "Cartesian perspectivalism." The latter is embodied in the ballistic perspective of the camera/eye moving toward a single depth point (think of the classic railroad vision produced from a camera mounted on the front of a train, or the penultimate scene of *Dr. Strangelove* with Major Kong riding the atomic bomb down to its target). Working from the philosophical premise of a fundamental link between perception and cognition, LaMarre argues that anime's lateral movement opens up a space of thought as a gap that unhinges our positions as seeing subjects and thus unhinges the received framework through which we view and interpret reality. Relative movement thus has the capacity to produce processes of thinking, relation, and being that are alternative to the instrumental rationality conveyed through Cartesian perspectivalism.

The point of the hypertext format is that it does not restrict the reader to a particular trajectory. On the contrary, it would seem to encourage a constant tacking back and forth between the available reading paths. Such shifts in perspective, it is important to note, are also shifts in scale between wide-angled overviews of the system that reveal its general features and close-ups that disclose increasingly complex detail. To borrow from the anthropologist of architecture Albena Yaneva, shifts enact rapid "jumps" in scale, allowing the reader to maintain two different experiences of a system simultaneously.[15] In addition, in reading this way the reader/player is constantly shifting back and forth between two modes of interaction that articulate competing desires of the network. On the one hand, the reader occupies a totalizing perspective of the system, a kind of technological eye in the sky that corresponds to aerial surveillance; on the other hand, the reader is provided micro-level, detailed examination of individuals in a way that produces a kind of ethnographic hyperintimacy. The overall effect is of a multiperspectival engagement with the experience of the subway that disrupts any attempt to impose a single determining logic or truth.

Ultimately, what holds the reader's/player's attention is not the story but rather the gamelike enjoyment of clicking from one link to the next in order to "zoom in" and "zoom out."[16] On a thematic level, the text offers only a suspended sense of satisfaction deriving from the anticipation that, with the next click, the reader/player will encounter something different. In sum, at least on a thematic level, the reader/player feels overwhelmed by the ordinary and predictable rather than compelled to reflect in a novel and critical manner on the processes and culture of commuting. But perhaps this is precisely what Inoue intended—a critique of the ordinary commuter and commuter experience.

Terminal Collectives

There is much in *99 Persons' Last Train* that is generalizable. The experience it simulates could be any city in the world, circa 1900 to 2000, where commuters return home exhausted from a grueling and tedious nine-to-five shift. Part of what lends *99 Persons' Last Train* a Tokyo-specific quality, however, is the nature of the last train and its collective. As I argued in the first chapters of this book, the commuter train system enfolds a constant tension—between control and emergence, routine and event, pattern and contingency—that is expressed in the gap between the planned *daiya* and the operational one. This tension realizes a certain kind of limit at the end of the day when service within Tokyo's commuter train network is brought to a temporary cessation, line by line, around midnight. Service then starts up again, line by line, beginning with the Chūō Line just before five in the morning. As the science-fiction writer and Tokyo-phile William Gibson observes, "Tokyo doesn't so much sleep as pause to allow crucial repairs to its infrastructure."[17] That time of pause transpires in accordance with the cessation of commuter-train service, ushering in an interval known as "after the last train" (*shūden no ato*).

Infrastructure repair is not the only activity that fills the time of "after the last train." While teams of maintenance and construction workers tend to the physical city under the glow of high-power halogen lighting arrays, other areas of the city open up to less-reputable if not quasi-illicit forms of labor and consumption. Notably, the area known as *Kabuki-chō* on the eastern side of Shinjuku Station becomes a lively scene of service-oriented consumption entertainment that includes everything from ordinary host and hostess snack bars to massage parlors, all-night manga and web cafés, and prostitution. Other notable, albeit perhaps less spectacular, scenes of "after the last train" consumption are found in the areas around Shibuya, Roppongi, and Azabu Stations, which are home to a number of dance clubs, upscale snack bars, and restaurants. The last train thus embodies a transitional moment in the structure of everyday life in Tokyo and a space of entanglement between two disparate articulations of solitudes. On the one hand, there are those who are trying to make it home after a hard day of work in order to recuperate just long enough to get through another day. On the other hand, there are those who are heading out to labor and consume in the marginalized interval of "after the last train." For the former, the last train is a terminal point of aspiration; it is the last chance for something to happen, some remarkable life-altering event, in an everyday otherwise marked by the numbing tedium of routine. For those among the consumers in the latter, the last train is the point of embarkation into a night world whose seemingly marginal quality imbues it with an uncertainty that translates into a sense of life-altering potentiality. And for those for whom "after the last train" constitutes the space and time of routine labor, the last train is merely a commute.

In situating his hypertext novel on the last train, Inoue taps into these tensions and into the different forms of collective that their solitudes articulate. At the same time, the last train in Inoue's story is not merely the last train of the day (*shūden*) but rather the *final* train (*saishū densha*). Whereas the last train marks a provisional stop, a pause within a recursive order, the *final* train carries a terminal weight, marking an absolute end. Indeed, for Inoue's characters this is not just another night on the last train of the day within the order and tedium of Tokyo life, but a final night that ends with nothing less than a "game over" in the form of a nuclear explosion in which solitudes coalesce into a single and conclusive but also irremediable collective. It is a spectacular event that transforms the last train into a life-changing experience, although perhaps not entirely in the way that the solitudes of the last train anticipated or desired.

The cause of the explosion is a malfunctioning robot, more specifically, an "autonomously mobile humanoid model assassination machine" with the designation "P13AX." The robot P13AX has been programmed to target the twenty-five-year-old granddaughter of an important politician, who is accompanied by a bodyguard whose presence she deems an intrusion on her independence. P13AX, we are told, is an unrivaled, state-of-the-art piece of computer technology with the capacity to perform its own system calibrations in the field and develop and execute its own plan toward fulfilling its mission. Things begin to go badly, however, when it follows its target into the subway. Immediately, its primary communication system malfunctions, which P13AX surmises might be the result of the electromagnetic waves emitted by the subway power lines. The malfunction is accompanied by "noise" (*noizu*) on the internal screen within its cognitive center. P13AX can determine that the noise is at 1654300 cycles but ironically can do nothing to fix it. It can only wait for a compensatory circuit to respond and override the amplifier. But because its entire primary communication system has gone into compulsory shutdown as a result of the initial malfunction, the compensatory circuit has been rendered inert. P13AX is doomed, it seems, by its own superlative logic—done in by an efficiency so efficient that it has realized an inverse effect. As a result of the impasse, the robot's head begins to oscillate, which appears to onlookers as if its head is trembling slightly. Abruptly, it begins to pound its head with its right arm in a steady, mechanized motion. The target onto whom P13AX is locked in a dead stare notices these strange motions and becomes increasingly frightened, thinking that the man (robot) is some kind of grotesque stalker. His/its uncanny behavior transforms an otherwise attractive figure into an object that stirs a sense of deep revulsion in the target.

In the meantime, increasingly cryptic and illogical system-status messages scroll across P13AX's internal screen as it goes into a state of shock:

Shock, Shock, Shock

"Please desist from striking this machine."

"Shock to a precision machine is liable to cause damage. Please do not strike or drop it."

"I visited him that day in his room."

"Do you vow to cherish and respect each other, through sickness and through health?"

"It's your wisdom tooth and it must be pulled. The root is rotted. You must have really neglected your jaw for it to become like this! Didn't it hurt? It's good that you came in today, you know."

"Shut up, asshole. Don't you know what time it is!"[18]

P13AX is at a loss to comprehend the error messages and is eventually overwhelmed by the illogical gibberish. As a last resort to carry out its mission, it moves close to its target and self-destructs with its nuclear power cell, causing an enormous explosion that leaves a gaping hole in the middle of Ginza Avenue. Game over.

99 Persons' Last Train ends the way most spectacular scenes of highly mediated mass death tend to end—with a perfunctory explanation in the newspaper:

Last night, at approximately 12:14 am.
Due to causes that are not clear, a large explosion occurred at the Ginza Station on the Eidan Subway Ginza Line, resulting in the collapse of a massive section of the Ginza main road. The collapse occurred in front of the Mitsukoshi Department Store, stretching sixty meters long and forty meters wide at the time when trains that had embarked from Shibuya Station and Asakusa Station were stopped at the Ginza Station. It is expected that there will be a great number of missing from among the passengers, train crew, and station employees. At this point, details concerning the damage are still unclear. Four hours after that accident, at 4:20 a.m., a special task force was established with the National and Metropolitan Police Forces, Fire and Disaster Management Agency, and Japanese Ministry of Defense to conduct an investigation. Due to continuing small-scale cave-ins since dawn at the scene of the accident, an evacuation order has been issued for all buildings in the area. According to the Japanese Ministry of Defense, there is a considerably high level of radiation detected at the scene, which calls for rigorous precautions. The task force cannot rule out at this time the possibility that this was an act of indiscriminate terror and are continuing to collect information from all related places.[19]

The reader gets the sense that Inoue arrived at the catastrophic ending for *99 Persons' Last Train* out of desperation to bring the ninety-nine different story threads together with some kind of closure. Yet the ending is also perfect in that it exemplifies the underlying weakness of the story, if not the central shortcoming of virtual practices in general: namely, the difficulty in constituting a lasting sense of collective. Indeed, the hypertext structure works well as a medium through which one can represent the multiplicity of narratives that emerge within the complex technological milieu of Tokyo's subway. It does not, however, remediate the space and time of the network in such a way as to offer an intuitive articulation of collective. Consequently, we are left with a gap, or more exactly a massive hole in the ground, in place of what might have been a collective remediation of the space-time of transit. Collectivity is only felt at the moment of its disappearance in the story.

Alternatively, we could read Inoue's ending as an attempt to convey the increasing sense of anxiety that urban Japan has felt over the course of a decade around matters of infrastructure and security. When Inoue began writing in 1996, many Japanese were still trying to make sense of the sarin-gas attack on a Tokyo subway train that had been carried out by members of the Aum Shinrikyo religious cult a year before. A few months before the subway attack, a massive earthquake destroyed much of the port city of Kobe, claiming thousands of lives. The tragedy in the latter event was compounded by the government's response, which demonstrated an unnerving level of incompetence. Two years before Inoue brought the hypertext novel to its conclusion, another large earthquake in central Japan caused the derailment of JR East's Jōetsu Shinkansen; miraculously this occurred without casualties. And finally, a year before Inoue finished the work, a JR West commuter train derailed outside Amagasaki near Osaka, killing 106 passengers plus the driver (an incident that I will discuss in chapter 6).

Densha Otoko

Moments of rupture in the paradoxical solitudes of the commuter train tend to be somewhat less spectacular and terminal than a nuclear explosion. Such moments are marked, we might even say, by a kind of gentle collective effervescence that invokes the sense of community and festival that Augé laments as being absent from the collective of commuter solitudes.[20] Train manner (chapter 2) is one way in which collective is displayed. There are also other, more spontaneous expressions of collectivity that take on the sense of a remarkable event. A friend told me of such a moment that happened during his morning commute from Tachikawa into the city on the Chūō Line. It was a Tuesday morning, and, as always, the train car was packed beyond capacity. On that particular morning, a young man wearing a large backpack was standing directly in front of the train door, facing the door such that his backpack jutted out into the crowded train behind him. When the train is packed during rush hours, it is customary for individuals to remove their backpacks and hold them in their hands against their chests, so as not to add more pressure to the already compressed environment. Not only had the young man failed to remove his backpack, he was wriggling his body back and forth so as to shove commuters standing around him and clear extra space for himself. My friend was not standing directly next to this man, but he was close enough to clearly feel and see what was happening. The young man's behavior was more than just a nuisance: it was creating added tension in the train car as his movements spread across the silent commuters cramped into the train like ripples from a rock dropped into a calm pond. Nevertheless, everyone silently bore the young man's ill-mannered behavior as best they could. Then at Yotsuya Station, something happened. At each previous station the young man had performed the maneuver, customary for riders standing at the door, of detraining momentarily to let others board, then boarding again himself at the end of the line so as to be standing again near the door. At Yotsuya Station, the man detrained momentarily as usual, then pressed his way back into the crowd in the door. This time, however, just as the train doors were about to close, in an unplanned expression of collective resolve a number of hands reached out and gently shoved the young man out the train doors and onto the platform. As the doors closed and the train pulled away, everyone on the train burst into spontaneous clapping applause.

Such moments become the stories that Tokyoites share with one another whenever the discussion, warmed by drinks and food, turns to the subject of commuting. The currency of such stories derives from their instantiation of a sense of collective that is comical and innocuous but also not without pedagogical rhetoric. The young man in the story gets what he deserves for his selfish behavior as commuters demonstrate a capacity to act not only in concert but also in the service of a certain justice of the collective in maintaining its order. The story articulates a line that the collective will not or at least should not allow to be passed, a line that compels commuters to momentarily redirect the energy they invest in maintaining their state of detachment from one another in order to solve a collective problem. There is something very satisfying about such moments when collective resolve is decisive and for the most part nonviolent. We might call those moments "enactments of collective morality."

Such instances of collective commuter resolve are, however, rare. More often than not, commuters turn away by turning inward, pretending not to see or hear the disturbance in the train car. An instance that produces a turning inward and away might not involve a major disturbance such as in my friend's account. It might be something simple, such as a commuter pretending to be asleep on the priority-seating bench while a pregnant woman or a child stands in front of them on unsteady feet. Discussions of these incidents, both major and minor, form the content of seemingly endless threads devoted to questions of train manner in various social-media platforms. Such instances are typically problematized as generational and exemplifying a collective moral failure as a result of pathological indifference at best or cultivated apathy at worst. Most often they are invoked as examples of the failures of Japanese youth—as if only youth are guilty of apathy. In these contexts, they are offered as proof of the thinning of human relationships caused by technological advance, particularly the spread of personal connectivity devices such as the *keitai*.

Such an incident of collective failure sparked the (supposedly true) events of *Densha otoko*. On the evening of March 14, 2004, the story goes, a twenty-two-year-old otaku posted an account on the Japanese-language subculture textboard site *2channel* of riding the train home from Akihabara and stepping up, when everyone else had turned away, to "save" a young, attractive woman from a drunk, middle-aged salaryman. Like everyone else in the train car, the otaku confesses, he initially pretended not to see the drunk staggering through the car, verbally abusing the women on the train. But when the man grabbed the chin of an attractive young woman seated across from him and yelled, "Women should just shut their mouths and let a man do as he pleases," the otaku mustered his courage and stood up to confront him. A brief verbal exchange and minor scuffle ensued, he wrote, but the incident was quickly brought to a conclusion by the arrival of a young salaryman, who helped subdue the drunk. Eventually, train attendants arrived as well. While the *2channel* thread participants commended the young man on his courage, they gently chided him for not asking the woman for a date. He has missed his once-in-a-lifetime opportunity, they stated, or so it would seem until two days later, when the otaku returns to the thread to report excitedly that the young woman has sent him a set of Hermès teacups to express her gratitude for his help that evening. Lacking experience regarding such matters—and, it seems, some degree of common sense—the otaku reaches out to the virtual strangers on *2channel* to seek advice. Inspired by the event, the thread participants rise to the occasion. They designate the young otaku "Densha otoko" (Train man) or just "Densha" (Train). The young woman they call "Hermès," on account of her gift, which they interpret as a sign of her refined character and taste. Coaching Densha in his courtship of Hermès, the thread participants hold council on how Densha should ask Hermès on a date when he calls to thank her for the teacups, and they prepare him for the date with a list of conversation topics and links to clothing stores, hairstylists, and restaurants, ultimately transforming him from an otaku into an adult, fashion-conscious Tokyo male. In return, Densha relays to the forum members reports on the relationship's progress and the events of each date over the course of the two-month courtship, drawing their envy as well as their continuous encouragement as he works up the courage to confess his love to Hermès.

As the discussion thread reached its conclusion with the love confession, one of the *2channel* participants began compiling the pages into six chapters, labeled "Missions," plus a short epilogue. A few months later, Shinchōsha picked it up and published the compiled thread in book form. Emphasizing the challenge to authorial conventions presented by the story's web-based bulletin-board origin, the story's author is listed as "Nakano Hitori," a play on words meaning "one among the group" and referring to both Densha and all of the anonymous *2channel* contributors.

According to the editor in charge of the *Densha otoko* project at Shinchōsha, another editor who often checked *2channel* came across the text by chance and suggested it for publication.[21] Shinchōsha's decision to publish the story of *Densha otoko* as a book made sense, as otherwise readers would not have been able to access the thread archive—the smartphone, and thus mobile access to the web, would not be introduced until the late 2000s. Carrying the text into the train simulated full web connectivity in ways that anticipated the introduction of the smartphone. Producing the *2channel* website in book form also lent the website a certain amount of mainstream legitimacy while making it available for individuals who might have hesitated to visit it due to its sketchy otaku reputation.

Apart from the buildup around the first date, the plot of the online and paper versions of *Densha otoko* is mostly unremarkable. The story basically involves Densha and Hermès going on a number of dates. Densha invests more attention and money into transforming himself from an unkempt otaku into a stylish young Tokyoite while he builds up the courage to confess his love for Hermès. Other than that, not much else happens. Densha and Hermès exchange a lot of *keitai* messages that get increasingly flirtatious; they discuss the film *The Matrix*; Hermès brings a friend on one of the dates, apparently to assess Densha; and on another occasion Densha visits Hermès's house for tea and scones. In the final "Mission," Densha helps Hermès purchase a personal computer, and they go for a drive along the coast. While parked, Densha confesses his love for Hermès and they kiss, thus supposedly sealing the relationship. Considering the lack of tension in the plot, it is understandable that the film and television versions of *Densha otoko* take certain liberties in changing the story. In the film, for example,

Densha loses his confidence at one point near the end and is ready to give up, but the virtual community rallies to revive his faith. Such changes lend much-needed tension to the story while bringing it in line with the representational norms of the cinematic medium. More importantly, they bring out certain latent themes in the original text version that speak to *Densha otoko*'s significance as an attempt to re-imagine the space-time of transit through digital connectivity. I want to focus here on those latent themes. In order to do so, however, it is necessary to briefly elaborate on the meaning and cultural significance of otaku and *2channel*.

The Otaku Medium

The novelty and popularity of *Densha otoko* had much to do with its *2channel* origin and the representation of otaku culture in *2channel*. Much has been written about otaku and otaku culture since the term gained international recognition in the 1990s. Although often trans-lated simply as "nerd," the term *otaku* is embedded within a vast so-cial discourse on Japanese youth that encompasses concerns over the alienating effects of technology and perceptions of a refusal among the younger generation to submit to conventional structures of labor and production.[22] Otaku are typically male, aged anywhere from the early teens to middle-age. They are often perceived as social misfits who har-bor a childish obsession with anime and manga, choosing these imag-ined worlds of fantastic futures, amazing technologies, exhilarating battles, and idealized, impossibly proportioned female heroines over reality. Otaku are seen as committing their life labor to these worlds, piously collecting related paraphernalia while embellishing, translat-ing, pirating, remixing, and circulating anime within a network of other otaku. Although certainly a generalization, there is a tendency within the Japanese media to equate otaku with the phenomenon of young Japanese who refuse to leave their rooms, so-called "shut-ins" (*hikikomori*). Although there is no definitive correlation between otaku and shut-ins, they are analogous in that otaku rejection of mainstream values of labor and production can be seen as an expression of the acute social withdrawal associated with shut-ins.[23] Otaku practices, it is often pointed out, predate the internet and the web. Scholars of otaku culture link the phenomenon's emergence with the advent of the video recorder, which allowed for obsessive watching, sharing, and modi-fying of animation.[24] The web simply allowed for the augmentation and amplification of these practices. When the self-professed otaku Nishimura Hiroyuki founded *2channel* in 1999, it became a generative infrastructure for otaku culture, providing a virtual home where otaku could discuss anything from new technology to anime, manga, and science-fiction gaming. What is more, *2channel's* textboard enabled the emergence of an otaku-specific mode of net communication characterized by the use of ASCII symbols for the creation of images and an insider lingo employing playfully distorted words and symbols.[25] For someone unfamiliar with the language, as I was when I first picked up the *Densha otoko* book, *2channel* feels like a foreign culture. Apparently, I was not the only one who felt this way, as Shinchōsha deemed it necessary to include a glossary of *2channel* terms in the text. So as not to ruin the sense of entering a different world for the non-*2channel* reader, the glossary was hidden under the book's dustcover.[26]

Unlike other text-based web forums such as bulletin board systems (BBS), a textboard allows for anonymity, as users can start or participate in a thread without registering or logging in. Postings are simply consecutively numbered. Adding another level of anonymity, *2channel* supposedly does not collect IP addresses.[27] These different mechanisms of anonymity are considered instrumental in *2channel's* earning a reputation for being something of a quasi-illicit virtual space where one can find private information on Japanese popular culture, celebrities, and politicians. In recent years, a number of individuals have exploited the site's publicity and anonymity to post their criminal intentions before embarking on acts of public violence.

Densha otoko did much to remediate the reputation of otaku, if not redeem the social value of the otaku from their association with criminal perversion. It made otaku culture and *2channel* not only accessible but also less threatening. The *2channel* otaku in *Densha otoko* come off as harmlessly immature but diligent and highly resourceful agents in an information society. Their threads are sometimes sophomoric, sometimes witty and endearing. At the same time, much has been said about *Densha otoko's* shortcomings. As the story transformed into a national phenomenon, weekly and monthly magazines featured articles scrutinizing its allure and its social implications.[28] Notably, a number of articles called attention to its unequivocal constitution of a male gaze and its corollary representation of the otaku ideal of women in the eternally forgiving maternal figure of Hermès. Others suggested that the story was perhaps an elaborate, government-initiated conspiracy to remedy the nation's falling birth rate by encouraging romantic relationships between the nation's most biologically unproductive populations: otaku men and unmarried working women in their late twenties and thirties, often disparagingly labeled in the media as "loser dog" (*make-inu*) women. Still others have critiqued the story's valorization of shallow consumerism and fashion, pointing to its failure to engage seriously with any real social issues.[29]

Networks of Encounter

On the one hand, as a tale of romance that embarks from an incident on the train, *Densha otoko* builds on the classic train-encounter trope in which strangers from different tracks of life, as it were, are brought together by an event on the train (often just a jolt of the vehicle) and set on a course toward some combination of romance, crime, or mystery.[30] Appearing throughout the literature and film of the twentieth century, the train-encounter trope invokes the notion of the train car as a paradigmatic space of twentieth-century urban modernity. Specifically, the train car figures as an intensified site of the urban crowd, enfolding its underlying tensions between structure and contingency, routine and event as forces of both positive possibility (romance, mystery) and negative risks (crime, accident).[31] On the other hand, if we follow the writer and literary critic Okazaki Takeshi, it is not the train-encounter trope but rather the more contemporary web- or net-encounter trope that drives the story of *Densha otoko*. After all, explains Okazaki, the allure of *Densha otoko* derives not from the romance between Densha and Hermès, but from the spontaneous emergence of a virtual community of netizens who flock to Densha's aid.[32] In urging us to exit the train and enter the "net," Okazaki's reading of *Densha otoko* aligns with a tendency that has been present in media and social theory since the advent of web communities, a tendency to abandon the social forms of mass-mediated society in favor of the emergent collectives of a postindustrial information society. Such a move renders themes of transportation secondary to motifs of communication as twentieth-century modernity is made to give way to twenty-first-century postmodernity. Or, as William Mazzarella argues in a critique of Michael Hardt and Antonio Negri's hugely popular treatise on the emancipatory potential of the postmodern information society, it encourages us to forsake the crowd as the bygone phenomenon of mass-mediated modernity's disciplinary enclosures and take up the fantasy of the "multitude" as an organic expression of unmediated and immanent collectivity.[33] The problem, Mazzarella's work suggests, is not one of crowds versus multitudes, as if either social form could successfully claim an inherent, substantive objectivity. Rather, to abandon the crowd is to succumb to a century or more of normative crowd theory, in which the crowd figures as an inherently regressive heteronomy, in favor of a popular and emergent network theory (multitude theory) promising emancipation through the "immaculate autonomy" of the networking multitude. All social forms, Mazzarella reminds us, are mediated—by which he means both immaterially and materially, theoretically and technologically. The trick, or rather the challenge, is to foster a "theory of social mediation" that invites us to think about the "emergent potentials of group energy" in the manner of their "dialectical coconstitution."[34]

What says "dialectical coconstitution" better than a story of "net encounter" titled *Train Man*? Indeed, the train does not give way to the web in *Densha otoko*. Rather, the story urges us to understand the multitude as a postmodern social form that can only be invoked from within the transitional everyday spaces of a capitalist industrial modernity. It encourages us to inhabit the machinic legacy of industrial modernity and its discursively constituted social forms in collectively enabling ways. In *Densha otoko*, the crowd is neither exterior to nor antithetical to the multitude; rather, the crowd is the multitude's inherent condition of possibility. Similarly, the story suggests that the material labor (commodity and physical) of industrial modernity is not exterior to the postmodern immaterial labor of the network. In short, *Densha otoko* asks us to understand that we have not left the train for the web, but rather that one is embedded in the other.

Immersions and Representation

Dialectical coconstitution transpires in *Densha otoko* in the synthesis the story generates between commuter crowd and web collective. The force of this synthesis rests on the corresponding ontologies of immersive technological mediation produced within the train and the web that make commuting and being online complementary experiences. To understand this, it is helpful to recall that for Schivelbusch, the novelty of the railroad as machine ensemble lay in its creation of a technologically mediated world. The machine ensemble interposed itself "physically and metaphorically" into the passenger's perception, Schivelbusch tells us, producing an immersive experience that radically altered the passenger's view of the world outside.[35] Looking out from the train window, the passenger saw not nature but rather an industrialized landscape mediated by the apparatus's "telegraph poles and wires" and linear speed.[36] In online communication, the keyboard, mouse, and monitor, experienced from within an immersive mediation that we call "virtual worlds," are akin to the telegraph and poles of Schivelbusch's railroad machine ensemble. The question then becomes how to represent the immersive experience of online communication in ways that not only are rhetorically interesting but also capture the energy and tension of virtual collectivity. Of course much is at stake in this capacity to represent if we recall the kind of progressive political potential that Walter Benjamin attached to the "revelatory optics" of film.[37] While certain common "fundamental instabilit[ies]" propagated by train and cinema—such as shock, mobilized vision, and simultaneity—made cinema the perfect immersive mode of representation for the train, cinema does not work quite as well for online communication.[38] Depicting a bunch of people typing away at their keyboards hardly makes for compelling cinema, let alone conveys the sense of collectivity that lends virtual communities their vivacity and seeming potential. Films like *The Matrix*, *Tron*, or even *Avatar* are attempts to capture that vivacity and potential through depictions of collective life in fantastic, immersive virtual realities. But such depictions also merely reiterate multitude theory in that they defer to the realization of collective as some kind of postmodern future separate from crowds.

In *Densha otoko*, the virtual collective and the commuter crowd do not belong to disparate modernities. Rather, they are simultaneously co-figuring and mutually allegorical social forms such that the commuter crowd serves as an avatar for the virtual collective, just as the commuter train network provides the mise-en-scène of *2channel* online communication. We see this most clearly toward the end of the film, when the thread participants restore Densha's confidence—a scene in neither the online nor text version of the story. Importantly, this is also the emotional apex of the cinematic version of *Densha otoko*, the moment when we feel the strength of the virtual collective's passion: in other words, this is the moment in which the collective is closest to a crowd. The scene begins with Densha returning home alone, in the pouring rain, looking dejected after confessing to Hermès that he feels they are too different to succeed as a couple. As Densha is sitting in front of his computer in his darkened room, the machine suddenly comes to life, its monitor displaying an open browser window within the *2channel* thread. A bright light begins to emanate from somewhere near the monitor, and the scene shifts abruptly to a virtual commuter-

FIGURE 4.5. Densha confronts the *2channel* crowd in the station.
Source: *Densha otoko*. Directed by Murakami Shōsuke. Tokyo: Toho Production Co., Ltd, 2005.

train station, which figures as the virtual matrix of the *2channel* thread. A dispirited Densha, whom we imagine as transported into this matrix, sits alone, slumped forward on a platform chair, looking down. When he eventually looks up, he finds a number of the thread participants standing across from him on the opposite platform.

"I'm sorry, I can't do it anymore," he says, and goes on to confess that he has no more courage to continue pursuing the romance with Hermès. The scene then shifts to individual thread participants, alone in front of their monitors, reading Densha's postings. As the thread participants begin to type their responses, the scene shifts back to the virtual train station, where they deliver their typed words to Densha face-to-face, as it were. The tacking back and forth between type appearing on computer screens and participants conveying their encouragement to Densha in the virtual station becomes more rapid as the passion of the other participants rises. Finally, the scene shifts to a rapid sequence of different participants writing, "Don't give up! Don't give up!" (*ganbare, ganbare*) on their screens, accompanied by the sound of thousands of fingers typing at computer keys that takes on the resonance of an approaching train. Only it is not a train: it is the commuter crowd, and when the camera takes us back to the virtual train station, we see Densha standing across from a platform crowded with *2channel* participants/commuters cheering him on. The virtual *2channel* collective, we are led to think, is the commuter crowd. More importantly, it is a crowd that has been remediated, in every sense of the word. It has been rendered through a different system of mediation, but it has also been made better, by which I mean transformed into a crowd that cares, a crowd that, unlike the commuter crowd in the story's initial train encounter, turns toward rather than turning away. But what, we must ask, is the nature of this care, and how is it invoked in relation to encounter?

The Ethos of the Multiplayer Game

If *Densha otoko* is the story of a virtual collective encounter (as Okazaki suggests), it is also the story of the initial missed collective encounter, when everyone except Densha turns away from the uncomfortable scene of harassment on the train. When the virtual *2channel* collective rallies to Densha's aid, first in helping him pursue a romance with Hermès and then again in propping up his courage, their actions remediate that initial missed encounter in the train. The ease and spontaneity with which the virtual community responds to Densha's appeal for help, moreover, stand in stark contrast to the collective failure of the commuter crowd to rise to Hermès's aid. But when the participants of the *2channel* group flock together to help Densha, neither manner nor a sense of moral responsibility drives them; rather, the ethos of the computer game, or "gaming," underlies the collective resolve.

Gaming is everywhere in *Densha otoko*. It is the core of an otaku ontology, and it is why Densha identifies himself in the first postings of the thread as "just a regular anime, game, Akihabara [*Akiba-kei*] otaku."[39] Gaming is behind the abundance of military metaphors in the story and is why the story is organized as a series of "missions" rather than chapters. Similarly, as in a computer game where players advance levels by completing increasingly difficult tasks, each successive mission in *Densha otoko* involves progressively difficult challenges that Densha overcomes by acquiring certain skills and knowledge. However, Densha is not playing an action game in the story: the challenge in action games tends to be surviving in an arena of individual competition, which makes action games a kind of simulation training for participation in the cutthroat environment of capitalist competition. By contrast, the gaming model in *Densha otoko* is the multiplayer cooperative game, like *Dragon Quest* (which, incidentally, is mentioned in the thread), in which players must move as a team through a virtual domain, gathering skills to overcome various challenges in order to advance to the next level. "We are all by your side" (*oretachi omae ni tsuite iru*), the *2channel* participants remind Densha. Indeed, Densha cannot advance to the next level without his team.

The ethos of the multiplayer game begins in *2channel* as a challenge to produce a subject thread, or *neta*, that will become dynamic and alive. As the Japanese literary critic Suzuki Atsufumi suggests, the objective is much like starting a volley to keep a ball in play between partners in a game involving an increasing number of participants.[40] The complex artwork constructed using computer coding text (JIS-Shift/ASCII artwork) that pervades the text version of the story is an extension of this same logic but adds an element of wit and a demonstration of technical expertise. With the first mission, the object of the game shifts from keeping the *neta* in play to the collective tasks of outfitting Densha, briefing him before his missions, and debriefing him when he returns. Each mission, moreover, transforms city space into an extension of "game space," an aspect that the film version visualizes by superimposing *2channel* text on the urban architecture. Each mission thus also figures as the *2channel* collective's foray into mainstream, normal, and normative urban space, which when read through the conflation of the gaming otaku as a shut-in (suffering from social-withdrawal syndrome) enacts the otaku as getting out into the world. At the same time, *Densha otoko* seems to suggest that the otaku is the new normative urban type, as the *2channel* otaku's command over the information network proves highly effective in allowing Densha to manage the obstacles and challenges of the city.

But the invocation of a gaming ethos in *Densha otoko* pushes us further. Computer games are not "failed representations of the world," writes Mackenzie Wark.[41] In fact we must consider the opposite: that the world is instead a failed representation of the computer game—the world, not the game, is broken. For Wark, game space is the corrective, a place in which we can learn critical thinking that can help us fix the world. Might game space then also be a place for learning how to care? Put differently, can otaku gaming provide a working alternative to the moral core considered lacking in a society in which everyone on the train turns away from the spectacle of harassment, turning instead inward into themselves? Can it cut through centuries of philosophical valorization of individual competition drawn from a reductive understanding of nature as the survival of the fittest and teach us to cooperate?

When the other passengers on the train with Densha turn away from the scene of harassment, they rehearse the sentiments from over a century of literature problematizing the social indifference of the urban crowd as yet another expression of the crowd's pathological irrationality.[42] In his seminal essay on the mentality of urban life, the sociologist and philosopher Georg Simmel identifies this indifference as the "blasé" attitude of the well-acclimated urbanite.[43] Writing at the turn of the twentieth century, Simmel attributed indifference to an expression of the inner aversion one develops toward strangers in the city in the process of forming a coherent sense of individuality, thus locking the individual and the urban crowd into an irresolvable but generative aporia that gets taken up ad nauseam in the literature and cinema of the era. To be in the city is to be torn between compassion for others and the desire to succeed as an autonomous individual in the competitive game that plays out on the political and economic topography of the urban landscape. The ethos of the multiplayer game overcomes this contradiction as it transforms the city into an extension of the game space where the underlying precept is self-organizing collectivity. The ethos urges one to turn toward, not away, not in accordance with any overarching philosophical imperative, but rather for the sake of realizing the crowd.

Turning(s) Away: The Other Side of *Densha Otoko*

The story of *Densha otoko*, I have argued, remediates the moment of collective moral failure when everyone in the train car turns away from the scene of harassment. It does so by folding the space of the train car together with the *2channel* community to articulate collective in the co-constitution of the commuter crowd and the virtual multitude. In the introduction to this chapter, I pointed to the way the cinematic trailer for the film *Densha otoko* allegorically stages this folding of commuter-train space and network space in its depiction of Densha and Hermès sharing a train car as texts from the *2channel* website stream by the train window. But if the trailer thus allegorically anticipates, to a certain extent, the remediation of the commuter crowd's turning away, it also stages a turning away from the commuter crowd and thus a rejection of the story's collective narrative. To recall, when Densha and Hermès communicate with one another via *keitai* from within the train in the trailer, they appear to have the entire train car to themselves. It turns out that the *2channel* crowd is present and part of this scene, except that it has been expelled from the central frame to the adjoining train car. We glimpse them only momentarily, peering in enviously

FIGURE 4.6. The banished crowd.
Source: *Densha otoko*. Directed by Murakami Shōsuke. Tokyo: Toho Production Co., Ltd, 2005.

through the windows at the end of the train car, where they have been relegated to being mere spectators of the story they have created.

In banishing the crowd from Densha and Hermès's train car, the representation enacts a disavowal of the packed train and thus a rejection of the condition of possibility of the story. As a rhetorical displacement of the crowd to the margins of the story, it reflects the thematic shift in the film from a story about the power of the virtual community with its ethos of the multiplayer game to a hackneyed tale of "pure love" (*junai*) between Densha and Hermès. As a pure-love story, *Densha otoko* becomes about the redemption of the *otaku*, who must learn to conform to the conventional desires of a heterosexual Tokyo adult male and spend his disposable income on appropriate services and commodities (stylish clothes, restaurants, and hairstylists) instead of infantile anime-related practices and paraphernalia. Similarly, the story becomes about the redemption of the "loser dog" (*make-inu*) unmarried woman in her thirties.

As a story about the normalization of the otaku and *make-inu*, *Densha otoku* becomes a tale about what Thomas LaMarre calls "media types": the endless stream of pathologized identities produced by Japanese media since the 1980s. For LaMarre, "media types" constitute evocations by Japan's media industry to indulge in an endless cycle of the customization of the self as a mode of relative movement within the onslaught of information flows that are constantly calling us to select and personalize consumption. As such, media types encourage a "turning away" from larger issues concerning the politics and aesthetics of technologies of collective life and a turning toward "little" nonpolitical questions concerning the stylization of the self within a network of culturally overdetermined consumer flows.[44] The commuter train, in this context, is no longer about the commuter crowd, with its inherent relation to mass production and urban modernity. Instead, the commuter train becomes a medium of niche consumption and lifestyles within an idealized vision of a twenty-first-century Japan as an exemplary information society.

The Rigged Game

What secures an individual's commitment to a system of their own exploitation? Classic Marxist theory answers this question by pointing to ideology, which it sees as distorting the exploitative reality of capitalist society's social conditions. To paraphrase Marx, individuals participate in a rigged system without explicit consciousness of their complicity in the mechanisms of their own enslavement. Marxist theorists defined this condition of unwitting complicity as deriving from a state of "false consciousness" that prevents a subject from recognizing the true nature of the alienating and exploitative social relations under capitalism. Accordingly, the task of the Marxist theorists was consciousness raising, which meant cultivating in the subject an analytic orientation toward social structure so as to lift the veil of ideological distortion and render intelligible the links between modes of social relation and means of production. Many thinkers have challenged the idea of false consciousness, arguing that individuals are far from being the naive subjects of ideology that the classic Marxist notion presumes. As Slavoj Žižek famously argues, it is more accurate to say that individuals are actually aware of what they are doing, but they just keep doing it anyway.[45] For Žižek, ideology works to secure an individual's devotion when it is incomplete and contradictory, not when it is a totalizing and seamless veil obscuring reality. Ideology, that is, does not colonize the mind completely. It leaves openings, gaps, as it were, for subjects to fill with their own fantasy construction in support of the irrational reality.

Žižek's thinking brings to mind Moroi Tomio, the central character in Ichikawa Kon's film *Man-in densha*. Moroi is a newly minted salaryman who struggles to succeed in a system that he knows is a rigged game. Even as his every attempt at success leaves him in more desperate conditions, Moroi insists on returning to the arena of labor with renewed vigor. Early in the film, Moroi offers an analogy comparing the socioeconomic circumstances that compel his struggle with the individual's subjection to the conditions of the packed train, telling a younger roommate just before he embarks on his new salaryman career, "You know, there's not a single seat we can occupy with hope in this country. But moping around won't get us anywhere closer to a seat on the packed train either. Society is rigged so that we have to work senselessly hard without good cause."[46] Moroi's interpretation of the system seems substantiated in the film's single packed-train scene, where we see him in a crowd of young salarymen pushing, shoving, and squirming to board a train that is packed so tight that the platform attendant, or "pusher," has to press the last man's back through the door with his foot.

But the film ultimately turns this interpretation on its head as we watch Moroi soldier on in life with comically energetic vigor despite a turn in fortunes that leaves him unemployed, too destitute to marry, and sharing a small shack in a rural area with his mother. As such, the packed train is transformed from a symbol of subjugation into an alle-

FIGURE 4.7. Stuffing in the last commuters.
Source: *Man-in densha*. Directed by Ichikawa Kon. Tokyo: Daiei Studios, 1957.

gory for the absurdity of individuals' fervent commitment to a system antithetical to their own material interests. The film's trailer distills this message with unveiled cynicism, declaring that Moroi's enthusiastic devotion to the rigged game makes him exemplary of the "contemporary new man" and the "youth who will develop tomorrow's Japan!" The "new man" is not a prisoner of false consciousness: he knows that capitalist society is a rigged game. Not only does he keep playing it, he plays zealously, devoting everything he has to the game. Only he does not use, as Žižek suggests, fantasy to fill the gaps in support of the irrational reality. What keeps him bound to the irrational reality is a sense of absolute despair that the system is too massive to change. For Ichikawa, the "new man's" fervent devotion to a seemingly hopeless condition makes him a more pathetic than comic figure.

Man-in densha was an early voice in the critique of Japan's postwar society that emerged among the nation's left-leaning intellectuals. Time and again this critique returned to the scene of the packed commuter train as the most vivid instantiation of the underlying contradictions at work in the nation's pursuit of high economic growth. Ichikawa's film participates in this intellectual movement by taking the nation's youth to task for their conscious devotion to a corrupt system, which could also be read as a thinly veiled criticism of citizens' support for Japan's imperial aspirations during World War II. Ichikawa wants the nation's postwar youth to understand that there is a choice, and that it is their devotion and despair, not the overdetermined structure of the system, that keeps them locked in precarious and exploitative circumstances.

Man-in densha was cinematic entertainment with pedagogical intent. It was an attempt at consciousness raising by mobilizing the packed train on a metaphoric level to convey the absurdity and irrationality of devotion to an exploitative system of postwar capitalist political economy. Unfortunately, Ichikawa's film was a complete failure. It performed poorly at the box office, and later in his life Ichikawa would declare it one of his least favorite works.[47] The nation's youth, it seemed, were not ready to embrace this message at a time when packed commuter trains figured as symbols of postwar economic recovery.

The *keitai* game *Days of Love and Labor* (*Ai to rōdō no hibi*) revisits *Man-in densha*'s theme of salaryman labor and the rigged game. However, it does so from *within* the train (rather than through the metaphor of the train) via a life-simulation game that transforms the train interior into a performative space of the rigged game. Whereas *Man-in densha* labored to politicize its audience by demonstrating the absurdity of devotion to the rigged game, *Days of Love and Labor* works from within the margin of indeterminacy. In so doing, it encourages us to think with the packed train toward a more complex response to the question of why individuals participate in a rigged game.

Days of Love and Labor

Days of Love and Labor was developed by the gaming-software company G-Mode as a "concept game" rather than an action game. While the latter tends to challenge the player's digital dexterity, a concept game engages the player through an idea or way of life. Accordingly, the overarching teloses (conquering, killing, making it to the next level) that provide the central motive in an action game are subordinate to a more abstract objective in a concept game. In *Days of Love and Labor*, the player is encouraged to reflect on the game's simulation experience, which the game's introductory page states as thus: "You work until retirement and then look back on your life. Your life in the game is over in thirty-one days. . . . Was it a happy life?"

Released in 2005, *Days of Love and Labor* quickly became one of the company's unexpected successes and was still one of its top-selling games when I downloaded it onto my *keitai* on the train in spring 2008.[48] The player's life in *Days of Love and Labor* consists of thirty-one work days during which the player experiences employment in a large company through a cute Blue Bear avatar. Each day takes about ten minutes to play. Blue Bear's job in the company is to roll a large ball to a prescribed location within a virtual environment while avoiding holes and other obstacles. With each successful completion of this task, the player earns wages that increase in accordance with the increasing complexity of the virtual work environment. Blue Bear's boss in the game is a Green Bear who appears at the start of every level to explain the new challenge. Green Bear also shows up to reproach Blue Bear for his incompetence when a player makes too many mistakes, such as allowing the ball to fall into a hole or off the platform. Such mistakes typically result in a demotion to a less difficult task and a corresponding wage decrease. At the conclusion of each workday, Blue Bear collects wages in Green Bear's office. With the earned wages, the player can purchase things (from a separate menu that pops up) such as a house, a car, a wife (a Pink Bear), kids, and so on. Wages can also be spent on entertainment. Regardless of the player's performance during the workday, Green Bear tends to invite Blue Bear to go drinking when

Blue Bear collects his daily wage. There is only one drinking venue, a yakitori place where the player can choose things like beer for fifty yen or chicken on skewers (*yakitori*) for twenty yen.

That Blue Bear is a salaryman during Japan's early postwar period of high economic growth is never made explicit in either the game summary or the minimal dialogue between characters. It is nevertheless an inescapable dimension of the game experience, conveyed through the language, affect, and ambiance of the game that work through multiple registers as recursive reiterations of salaryman norms. The effect is that players do not simply experience the life of a salaryman through their Blue Bear avatar in *Days of Love and Labor*, they become salarymen in ways that recall Judith Butler's discussion of gender performativity.[49] When Butler argued that gender is performative, she reconceptualized decades of social theory that privileged language as the medium of discursive power. Such theory is exemplified in Louis Althusser's seminal critique of state power as interpellating subjects in the moment the individual responds to the hail of state authority.[50] For Butler, simple cognitive recognition of state authority was not enough to explain the work of power, just as it was not enough to say that gender is a discursive construct socially constituted through normative articulations of sexuality. One had to pay attention, Butler argued, to the processes by which discourse materializes in the body. For Butler, this meant exploring the way gender literally becomes embodied by tracking its transposition from a normative social precept to a mode of being that risks foreclosing the potential multiplicity of becoming. Performance, for Butler, was the site of analytic focus, where gender permeates the material body through reiterations of gestures, speech, and stance that materialize at a corporeal level in sedimented patterns of being. Through Butler's thinking, discourse thus exceeds its linguistic confines to become a material force with the power to shape actual bodies, not just thought. If bodies thus matter for Butler, they matter because they are processes with potentials and limitations inseparable from their immediate cultural milieus and historical context.

By undertaking the labor of pushing a ball around within a virtual environment, the player finds him- or herself moving in a performative register toward becoming a salaryman. Lacking any readily intelligible meaning, the labor is absolutely alienating, reducing one to a mere cog in the machine incapable of cultivating a sense of self-worth independent of the company.[51] Repetition is key. Each day of labor becomes a reiteration of normative salaryman values while the player begins to look expectantly forward to the appearance by Green Bear, whose random reprimand or praise becomes the only metric of self-worth. Green Bear's invitations become equally difficult to refuse as the player begins to feel increasingly embedded in the company life. Equally important for its performative dimension, *Days of Love and Labor* is not just played *on* the train, but *with* the train and the recursive rhythms of the commuting experience. On a most basic level, the repetitive practice of commuting is analogous with the repetitive labor the player performs as Blue Bear. The virtual space of the game and the space of the train car come to mirror one another such that the train becomes the game and the game becomes the train. The spaces mutually reinforce the element of performance, transforming virtual and corporeal repetition into embodied reiterations of normative salaryman values.

Despite G-Mode's designation of *Days of Love and Labor* as a concept game, its concept is closer to what Lev Manovitch calls the "narrative shell" of an action game.[52] The narrative shell is the story frame for the game algorithm. It dresses the functional and recursive logic of the game with symbolically meaningful content, such as "aliens are attacking the earth and you must find and destroy their lead ship." As Manovitch suggests, the narrative shell is often unimportant to the player. It merely provides a heuristic for the player who becomes focused on discovering the underlying logic behind the game articulated in the algorithm. Thus in some cases, gaming can be seen as an introduction to simple Marxist theory in that discovering the algorithm is methodologically commensurate with revealing the social and organizational logic of the base structure of capitalist political economy.

What the player quickly learns is that *Days of Love and Labor* is a rigged game. No matter how well one works or strategizes to accumulate savings, the game saddles the player with enormous debt compounded by some combination of divorce and demotion. My own experience playing the game illustrates this. After about a week of playing during my commute, I had finally mastered Blue Bear's work and accumulated 2,000 yen in virtual money. Having gained some confidence, I decided to browse the "Purchase Something" list, where for 200 yen I bought a Pink Bear wife, who told me right away that she loves me (*aishiteru wa, anata*). When I returned home the next day after earning 700 yen at work, my wife welcomed me home with a customary greeting of "Welcome home, you've worked hard today" (*oshigoto gokurō sama, anata*). She then asked if we could buy a car for ¥1,000 yen, and I agreed, thinking that I could recover the sum quickly in a few days of work. But that was only the beginning of expenses that would quickly accumulate. Upon returning home from work the next day, I was informed by my wife that while I was at work she made some purchases. I learned that I owned, among other things, a television and a bicycle, and that I also owed ¥7,000 yen to a loan agency. The news was distressing. Determined to pay back my loan in a timely and responsible manner so as to avoid incurring fines, I worked consecutive days without rest, each day refusing as well Green Bear's invitations to go drinking. When I had collected almost ¥4,000 yen, my wife suddenly informed me that she had been lonely with me always at work and so had decided to divorce me and return to her parents' home. Soon after that, a representative from the loan agency—another Green Bear—appeared at my home and took away my television, car, bicycle, and furniture as payment on my overdue loan. I finished my life working to pay off the rest of my loans. Although I played countless times and my performance improved, I never managed to finish my thirty-one-day life without severe debt compounded by divorce or demotion. Other players, I learned from a website devoted to discussing the game, had similar experiences:

128: In the end it was just a hard life. Every day being chased by the loan collector and having things taken away. I wish I could have at least saved my wife and kids, I'm sorry . . . sorry . . .

154: Not used to the work, I kept making mistakes and getting reprimanded by my superior day after day. Finally, just when I got promoted, I learned that my wife had taken out a loan. For the sake of my wife, I put my spirit into my work. But I ended up getting demoted for consecutive misses. Then the loan collector barged in at my home and took all the things I'd bought when I was single, my car, bicycle, cell phone. Finally, I figured out the work, and as I worked to pay back my loan, I neared retirement. Just as I thought I'd paid back my loan, I discovered a new loan. Approaching retirement. Nothing but an absolutely bitter life. Zero points.[53]

Days of Love and Labor makes no effort to conceal the algorithm in its narrative shell. To play the game is to struggle to realize happiness through labor only to be confronted with the impossibility of succeeding. In other words, the rigged game is the game. One might assume from this, as I did, that *Days of Love and Labor* was meant as social critique. It was not. According to the game developers, a team of four men in their early thirties and early forties, whom I interviewed in the summer of 2010 at the G-Mode company headquarters in Shibuya, *Days of Love and Labor* was never intended to convey critique of any *contemporary* significance. The game was intended rather to provide an enjoyable parody of the 1960s salaryman life experience for today's young Japanese. "It is about then, not now," the game designers emphasized. In order to ensure the simulation would remain light-hearted and unprovocative in a critical sense, they removed all complexity from the narrative shell. Similarly, it was decided to exploit minimal, low-definition and flat graphics instead of high definition and detailed graphics. The latter, explained the game designers, puts too much onus on the player's concentration and subtracts from the transient, suspended sense of self, or what they called the commuter's "unfocused feeling" (*bonyari shita kanji*), that makes the commute conducive for flights of fantasy. (This last design strategy around the graphics has significant effects for the play experience within the train, to which I return below.) *Days of Love and Labor* was thus not supposed to pose difficult questions. Its narrative premise and algorithm were meant to converge to produce the simple idea that the salaryman of the 1960s was a prisoner of a culturally overdetermined economic system from which contemporary work is free. But of course, the best technologies exceed their programming. *Days of Love and Labor* is interesting not as a (mindless) game about 1960s salaryman labor but rather for the way it asks and responds to the paradox of the rigged game from within the space of the train.

The Train and the "Zone"

As a rigged game played within conditions of immersive technological mediation, *Days of Love and Labor* shares an affinity with the experience that Natasha Schüll depicts in her ethnography of machine gambling in Las Vegas. For Schüll, the question machine gambling posed was, similarly, Why do players persist despite fully understanding the game is rigged? Winning, she finds, is not their objective. Rather, players gamble in order to enter "the zone," which Schüll describes as a transitory state of consciousness in "which time, space, and social identity are suspended in the mechanical rhythm" of machinic intimacy.[54] For Schüll, the zone speaks to an underlying existential anxiety that has become part of the technological condition of contemporary life. Providing a sense of intimacy and predictability, it becomes something of a detached space of machinic comfort through which individuals can "manage their affect states" amid an increasing sense of uncertainty in the world.[55] The novelty of Schüll's argument, however, lies in showing how, despite being experienced as a kind of organic and serendipitously realized state of mind, the zone is actually the effect of intense design in which minute attention is paid to every element of the environmental ambiance, from the sounds of the machines to the lighting, the textures, and the field of vision. Similarly, every aspect of the gambler's feedback relation with the machine is carefully designed, from the look and feel of the console buttons to the cards that are just barely visible at the edge of the screen on an unsuccessful turn. The paradox is that while casino design aims for a totalizing, immersive technological environment, it crafts tactical gaps in the gambling-machine cycle and environment to interact with and draw in the gambler's psyche.[56] The trick of casino design, in other words, is the trick of the rigged game—the creation of a totalizing immersive design that nevertheless remains undetermined.

It is not difficult to draw a comparison between the commuter train and the Las Vegas casino as Schüll describes it. Commuter-train cars, like casinos, are engineered spaces of immersive mediation that lend themselves to a zone effect, or what the G-Mode game designers described as an "unfocused feeling."[57] But the affective organization and subsequent quality of their respective zones could not be more different. If, for the casino gambler, "the zone" is realized in rhythmic and isolated intimacy with the machine, for the commuter the zone lies in realizing a point of perfect equilibrium between attention and inattention, between concern with and indifference to the crushing intimacy of commuter bodies. For the commuter in the packed train, this equilibrium is akin to a balancing stance in yoga, in which one achieves an absolute minimum of energy expenditure through intense physical focus and active mental concentration. The effect is of being proprioceptively suspended between connection and disconnection, between intimacy and alienation within the crushing pressure of commuter bodies.[58] The *keitai* game intersects perfectly with this point of equilibrium on account of the dynamic between attention and distraction that defines the character of its interface. As the anthropologist Jun Mizukawa suggests in her analysis of the temporality of the *keitai* novel, reading on *keitai* typically happens while in motion (walking, biking, riding on a bus or train) or within infrastructures of mobility (train stations, bus stations, and so on).[59] In other words, it happens "in conjunction with all the other reading that must be undertaken beyond our small cellphone screens" when we must "look up, however briefly, to make sure not only that we are not missing our station, bus stop or destination but also not running into people, city structure, or objects that surround us."[60] Such moments of looking up from the screen, Mizukawa argues, are not distractions from the text. Rather, they are part of a modality she defines as "reading on the go," whereby the atmospherics of the city, with its varied tempos and noises, become enfolded into the text when the reader's eyes return to the screen. "Reading on the go" animates and enriches the text, lending it a vitality that otherwise tends to be absent in the writing. Although Mizukawa deals with reading on the *keitai*, her argument is equally relevant for *keitai* gaming. The surrounding milieu is not a distraction: taking on a logic of supplementation, it completes the gaps in the game. *Days of Love and Labor*'s low-definition graphics incorporate this modality of interaction. In contrast to high-definition graphics, which become a vortex of stimuli that capture and contain the player's attention, the incomplete and unfinished quality of low-definition graphics performs as lacuna, inviting in the dynamism of the surrounding world.

In *Days of Love and Labor* the surrounding milieu of the train car animates the game's central desire. Put differently, the impossibility of interaction within the crushing intimacy of the train car produces a longing for the simulated intimacy of human interaction within the game. Although *Days of Love and Labor* begins as a game that promises simulated employment, it is the promise of simulated intimacy that takes over to define the gaming experience and determine the player's response to the question, "Was it a happy life?" Simulated intimacy in the game begins simply, with Green Bear's invitations to go drinking at the end of the day. But those brief excursions with Green Bear, in which there is only a menu of items to choose from and no exchange of dialogue, can only prove dissatisfying. Thus the player is compelled to look beyond the confines of work for a different and more complex kind of interaction: enter the wife and children. The wife's immediate declaration of her love fills the space left by the unsatisfying excursion for drinks with Green Bear. The wife's subsequent idiomatic expressions ("Welcome home, you've worked hard today") lend depth to that space by conveying a sense of routine and familiarity.

But intimacy in *Days of Love and Labor* is quickly transformed into responsibility when the wife begins borrowing money for purchases. Sustaining and providing for the relationships one forms with the wife and then children become the player's objectives and the reasons why the player returns each day with renewed dedication to work. As a player wrote in comment 154 of the Logsoku.com thread for this game,

"For the sake of my wife, I put my spirit into my work."[61] Nothing compels players to put their "spirit" into their labor and accept responsibility, other than the desire for continued simulated intimacy. Love, not labor, thus becomes the game's motivating factor.

However, the negative portrayal of the wife makes it difficult not to read *Days of Love and Labor* as misogynistic.[62] Not only does the wife lack any positive qualities, but also her true function in the game seems purely malicious: burdening Blue Bear with debt and then abandoning him when he is in the midst of struggling to pay it back. In other words, the wife is a ruse. She promises to fill the gap with a sense of intimate connection but delivers instead only debt, divorce, and despair. Via the wife, the meaning of love in the game is reduced to a mechanism of exploitation and alienation more powerful than wage labor. On the other hand, the game also invites a different interpretation. The sentiment "I wish I could have at least saved my wife and kids, I'm sorry . . . ," conveyed in comment 128 of the Logsoku thread, suggests that the game does not necessarily elicit misogynist resentment.[63] Far from being disparaging, such comments suggest that the wife is a victim of a system in which one's ability to care for someone is made contingent on one's capacity to labor. The problem is thus not specifically the wife, but the quality of collective relations.

By producing a player's commitment to the rigged game through the desire for intimate connection, *Days of Love and Labor* offers a richer and more empathetic critique of an individual's position within capitalist society than Ichikawa's film *Man-in densha*. In the latter, the conscious devotion of the "new man" to a rigged system expresses not only the fundamental, irrational rationality of capitalist society but also the new man's failure to struggle as a rational being to overcome those irrational circumstances.[64] The new man, Ichikawa tells us, may have a high threshold for despair, but he has a weak will. He thus holds Japanese youth responsible, presenting their irrational surrender to the despair of the rigged game as the source of society's continuing irrationality. Ultimately, Ichikawa wants us to understand that the conditions that allowed for the violence of imperial Japan still exist.

By contrast, *Days of Love and Labor* asks us to think of devotion to the rigged system not as the result of weakness and irrationality but rather, in far more realistic terms, as an expression of the sticky reality of emotional entanglement. Desire for attachment—attachment to attachment—perpetuates one's participation in the rigged game. Such desire for attachment cannot be called "cruel optimism," because it lacks the necessary illusion of hope and the accompanying unsubstantiated sense of confidence that characterize the kinds of attachment to fantasies of the good life that ultimately prove self-destructive.[65] There is no illusion and no hope in *Days of Love and Labor*. The player knows that the game will end in some debilitating combination of debt, demotion, and divorce, just as Moroi knows in *Man-in densha* that there is no hope to be found in a seat on the packed commuter train. The view of reality from within the packed train in *Man-in densha* as well as in *Days of Love and Labor* is unobstructed by either ideology or fantasy. But what sends the player back to work in the latter—what brings the commuter back to the train every day—is the sobering insight that there is no individual outside of individuation.[66] One cannot exist alone. It takes a collective to make and sustain a person. But collectives need more than connections in order to further that individuation. In order for one to be able to answer the question "Was it a happy life?" with something more definitive than "yes, maybe, I don't know," collectives require attachments in the form of relations bound in a reciprocity of mutual becoming. The absence of a condition of possibility for such attachments within the packed train is the initial impulse that sends the commuter to labor as Blue Bear in *Days of Love and Labor*.

Only from within the collective of the packed train is *Days of Love and Labor* able to bring forth such thought about the rigged game. Via the crushing intimacy of bodies within the packed train, the simulated experience of laboring in the rigged game successfully calls into question the quality of the collective elicited within Tokyo's commuter train network. Put differently, the game produces a recursive feedback dynamic between the player and the packed train that carries the potential to generate a critical perspective of the technicity of the network. Whereas Ichikawa mobilizes cinema to deploy the packed train as an allegory for the irrationality of capitalist society, *Days of Love and Labor* elicits thinking with the packed train to question the potential for becoming that is afforded within its matrix of human and machine relations.

Conclusion: Closing a Gap

In recent years, posters began appearing in trains and train stations throughout Tokyo asking commuters to refrain from using their smartphones while walking. Although the content varies in these posters, the general message is that smartphones are black holes for human attention that put commuters at risk of colliding with objects, walking off the station platform onto the tracks, falling down stairs, or any number of unfortunate incidents likely to befall the commuter who is unable to embody the dynamic of attention and inattention necessary for inhabiting the commuter train network. In contrast to the *keitai*, the smartphone has become an addictive means for producing an isolating intimacy within the collective of the commuter train network, making it similar to the gambling machines in Las Vegas casinos.

The relationship between the web and the train network that I have elaborated throughout this chapter belongs to a particular moment of in-between when the *keitai* was transformed from an advanced text-messaging device into a medium of limited internet connectivity. That relationship, and the particular possibilities for remediating the commuter train's margin of indeterminacy that relationship invoked, were brought to a close with the introduction of the smartphone with full internet access. This is not to say that the latter does not offer ways for thinking and experiencing the commuter train network differently. Rather, the kind of attention it engenders and the character of its interaction with the space and time of the commuter train network produce something different from what I have presented here. It is perhaps too early in the technology's evolution to say what that something is. In the evolution of the relationship between the web and the commuter train, the gaps between the two, which were embodied in the *keitai* with its limited degree of interface, were crucial to the *keitai* moment. Insofar as both commuter train and web produce immersive technological mediations, those gaps suggested that the web and the *keitai* were not seamless, symbiotic systems but rather specific expressions of technicity inseparable from the spatiotemporal organization and functionality of each ensemble. This disparate technicity and the subsequent distinct quality of collective that it entailed allowed computer gaming to articulate an intervention in the space and time of commuting in the form of the question, "Can the commuter train teach us to care?" That question derives not from the particular technicity of computer gaming; instead, it emerges in the gap of interaction between the commuter train network and the web. Each instantiation of thinking the train with the web presented in this chapter asks and answers that question differently. But they all do so in ways that come back to examine the quality of relations of the collective of Tokyo's commuter train network.

Notes

Some of the material in this chapter has also appeared in Michael Fisch, *"Days of Love and Labor*: Remediating the Logic of Labor and Debt in Contemporary Japan," *positions: asia critique* 23, no. 3 (2015): 463–86.

1. Kirby, *Parallel Tracks*; Schivelbusch, *The Railway Journey.*
2. The term *otaku*, which loosely translates as "nerd" or "geek," is discussed in depth later in the chapter.
3. Frasca, "SIMULATION 101."
4. Manovich, *The Language of New Media.*
5. Pickering, *The Mangle of Practice.*
6. Augé, *In the Metro.*
7. Augé, *Non-Places*, 120.
8. Ibid., 94.
9. Inoue, Interview with a Shinchōsha editor.
10. Inoue, *Kyūjūkyū nin no saishū densha* (*99 Persons' Last Train*).
11. Gitelman, *Scripts, Grooves, and Writing Machines*, 222. It is worth noting that Inoue was not the only one to experiment with hypertext and subway fiction. In the same year that Inoue began uploading sections of *99 Persons' Last Train*, the Canadian science-fiction and fantasy writer Geoff Ryman launched a hypertext novel on the web titled *253* that follows the lives of 253 commuters on the London Underground; Ryman, *253: The Print Remix*. Nothing in Inoue's interview suggests he knew of Ryman's work when he started *99 Persons' Last Train*. As might be expected, a number of similarities occur in Ryman's and Inoue's works. Each allows the reader to move according to character perspective or system structure, and each ultimately progresses toward a major catastrophe. However, Ryman's work is less complex in that each of the characters is presented according to a set format of outward appearance, inside information, and what the character is doing or thinking.
12. See Tom Boellstorff's excellent ethnography of *Second Life, Coming of Age in Second Life.*
13. LaMarre, *The Anime Machine*, 103–9.
14. An example LaMarre provides is the sense of movement imparted when, sitting on a stationary train, one observes through the window another train beginning to move. Although the train one is in does not actually move, the lateral movement of the other train viewed from one's train window creates a sensation of movement.
15. Yaneva, "Scaling Up and Down."
16. Ibid.
17. Gibson, *Pattern Recognition*, 140.
18. Inoue, *Kyūjūkyū nin no saishū densha*. My translation.
19. Ibid.

20. Augé writes that the Paris Métro is "collectivity without festival and solitude without isolation"; Augé, *In the Metro*, 30.

21. Shinchōsha editor, interview by the author, February 2005, at the company's main office in Tokyo.

22. See Ito, Okabe, and Tsuji, *Fandom Unbound.*

23. Saito and Angles, *Hikikomori*; Horiguchi, "Hikikomori."

24. LaMarre, "An Introduction to Otaku Movement."

25. Suzuki, *Densha Otoko wa dare nanoka.* See also Fisch, "War by Metaphor in *Densha Otoko.*"

26. According to the Shinchōsha editor, searching for a word in the hidden glossary would be like performing a *kossori kensaku* (secret web search).

27. Lisa Katayama, "Meet Hiroyuki Nishimura, the Bad Boy of the Japanese Web," *Wired Magazine*, May 19, 2008. Katayama tells how Nishimura established the site while studying psychology at the University of Central Arkansas. The site's servers are in San Francisco, where they remain safely out of reach should the Japanese government ever decide to act on the numerous lawsuits against the site or on its threat to curtail freedom of web expression.

28. *"Densha otoko* ga būmu ni" [*Densha Otoko* becomes a boom], *Spa Weekly Magazine*, December 13, 2005, 91; Yohei Fukui, "Densha otoko ni naritai bokutachi" [We want to become Densha Otoko], *Aera Weekly Magazine*, February 2, 2005, 12–17. See also Yohei Fukui, *"Densha otoko* no iya onna ron" [The *Densha Otoko* discourse on undesirable women], *Aera Weekly Magazine*, October 17, 2005, 16–19.

29. Uchiyama, Hiroki, *"Densha otoko* hamaru onna no ren-aijukudo" [The level of romantic maturity of women who fall for *Densha otoko*], *Aera Weekly Magazine*, November 22, 2004, 29–30.

30. Fisch, "War by Metaphor in *Densha otoko.*"

31. Kirby, *Parallel Tracks.*

32. The real story in *Densha otoko*, Okazaki suggests, is about the virtual community of netizens who come together to aid Densha. The event on the train, he argues, is only significant insofar as it provides the pretext for that plot. Okazaki, "Bestoseraa: Shinsatsushitsu."

33. See Mazzarella's critique of Hardt and Negri's *Multitude* in Mazzarella, "The Myth of the Multitude, or, Who's Afraid of the Crowd?"

34. Mazzarella, "The Myth of the Multitude," 715–16.

35. Schivelbusch, *The Railway Journey*, 31.

36. Ibid.

37. I draw the term "revelatory optics" from Flaig and Groo, *New Silent Cinema*, in my reading here of Walter Benjamin's "The Work of Art in the Age of Mechanical Reproduction." See Flaig and Groo, 168.

38. Kirby, *Parallel Tracks*, 3.

39. Nakano, *Densha otoko*, 12.

40. Suzuki, *Densha Otoko wa dare nanoka.*

41. Wark, *Gamer Theory*, para. 021. (Wark's book has paragraph numbers rather than page numbers.)
42. See Mazzarella's critique of literature on the crowd in "The Myth of the Multitude."
43. Simmel, "The Metropolis and Mental Life," 14.
44. LaMarre, *The Anime Machine*, 107–9.
45. Žižek, *For They Know Not What They Do*.
46. Ichikawa, *Man-in densha*, 8:00–8:23.
47. Ichikawa and Mori, *Ichikawa Kon no eigatachi*.
48. Fisch, *"Days of Love and Labor."*
49. Butler, *Bodies That Matter*. My reading of Butler's thesis is greatly informed by Adrian Mackenzie's highly insightful interpretation in *Transductions: Bodies and Machines at Speed*.
50. Althusser, *On the Reproduction of Capitalism*.
51. My interpretation of the game here is based on the game designers' description of its intended effect during my interview with them in the summer of 2010.
52. Manovich, *The Language of New Media*.
53. 2channel, accessed May 1, 2017, www.logsoku.com/r/appli/1115218839.
54. Schüll, *Addiction by Design*, 13.
55. Ibid, 47.
56. Gaps take the form of corners and nooks sheltered from the central flow of people, or a glimpse of a winning combination just barely visible at the edge of a losing turn. Gaps are spaces that beckon as ontological openings.
57. If we follow Schivelbusch, the train car is the original scene of immersive mediation, where human perception and experience of the world was first subjected entirely to the speed and architecture of the machine ensemble.
58. The term *connection and disconnection* refers to Anne Allison's description of commuter sociality as "disconnected connectedness" in *Millennial Monsters*, 72. Similarly, the term *intimacy and alienation* is drawn from James Fujii's parsing of the commuter experience as one of "intimate alienation" in "Intimate Alienation: Japanese Urban Rail and the Commodification of Urban Subjects."
59. Mizukawa, "Reading 'on the Go.'"
60. Ibid., 80–81.
61. 2channel, www.logsoku.com/r/appli/1115218839.
62. As I have written elsewhere, I was surprised to learn that G-Mode designed *Days of Love and Labor* with the intention to market it to young women; Fisch, *"Days of Love and Labor."* The rationale behind this was that they thought the game would feel "too real" for young male players.
63. 2channel, www.logsoku.com/r/appli/1115218839.
64. I use the term *irrational rationality* in reference to Max Weber's critique of modern bureaucracy. The Weberian critique in the film is implicit. Weber's work was popular among Japanese university students at the time.

Moreover, that Ichikawa was familiar with Weber's thesis is clear from another of his films from around the same time, *The Punishment Room* (*Shokei no heya*) (1956) in which he depicts students debating Weber's texts.

65. Berlant, *Cruel Optimism*.

66. The idea that the genesis of the individual is in the process of individuation is a fundamental concept in Simondon's work. See Simondon, "The Genesis of the Individual."

FIVE

Forty-Four Minutes

If you are jumping from a platform choose a station where the express train does not stop as the chance of fatality is much lower if the train is decelerating. If you misjudge the timing you might bounce off the front of the train or jump too far and end up on the other side of the tracks. So take your time. As long as the train is within a hundred meters of the station, it is not going to be able to stop in time with the emergency brake. Also, be careful since there have been incidents in which briefcases or other items have flown back and hit someone on the platform.

ADVICE FOR HOW TO COMMIT SUICIDE BY JUMPING IN FRONT OF A TRAIN THAT WAS POSTED ON A WEBSITE DEVOTED TO TECHNIQUES FOR SUICIDE.[1]

It is quite possible that the man who jumped in front of the Tokyo-bound Chūō rapid train at Higashi-Koganei Station at 7:42 a.m. during the morning rush on August 19, 2004, read this advice on the website. His black nylon briefcase, a fashionable sporty type with a padded section for a notebook computer, lay abandoned at the far end of the platform where the train had entered the station. Although the train was decelerating as it entered the station, it was still moving fast at the time of impact. The driver had not managed to sound the horn, no doubt devoting all his efforts instead to braking before the body could become impossibly tangled in the wheels. It all happened very fast. Aside from a few reserved gasps of astonishment from commuters waiting on the platform, there was not a great deal of commotion. Because of the considerable rush-hour crowd on the platform, those toward the middle and the far end could not immediately see what had happened. They figured it out soon enough, however, from the splotches of blood on the front of the train as it came to a halt one-quarter of the way into the station. Most turned away, murmuring only *"mata ka!?"* (again!?) in a tone that conveyed a mix of exasperation and bewilderment. Others expressed revulsion—"disgusting!" *(iya, kimochi warui!)*—before flipping open their *keitai* to inform whoever expected their arrival that, once again, they would be late. The train, in the meantime, remained halted halfway into the station with its doors shut and its passengers standing tightly pressed up against the windows, visibly irritated but quiet. Within moments, the stationmaster delivered the first of what would be many announcements over the station's public-address system: "At present there has been a *jinshin jiko* [(human) body accident] on track 1 in Higashi-Koganei Station."

Commuter train suicides have been a facet of urban life in Japan since the early decades of the twentieth century when commuter trains became an infrastructural mainstay throughout the nation. The term *jinshin jiko*, however, is relatively new. Composed of the Kanji characters for "(human) body" and "accident," the four-character compound is an abbreviation of *jinshin shōgai jiko*, which means "an accident resulting in bodily injury." This is a general classification for any incident involving human injury or death as a result of a mishap with a vehicle.[2] In the context of the railroad, the term encompasses four possible circumstances of death or injury: suicide *(tobikomi jisatsu)*, entering the track area *(tachiiri)*, falling from the platform *(tenraku)*, and unintended contact with a train entering the station *(sesshoku)*. Referring to all four possibilities, the term *jinshin jiko* renders the cause behind the body on the tracks indeterminate.[3] As this categorization was introduced for administrative and insurance purposes, most commuters are either unaware of it or at least unfamiliar with its specifics.[4] For the average commuter, an announcement of *jinshin jiko* tends to mean one thing: a commuter train suicide.

Although officially introduced in the late 1980s in conjunction with the privatization and reorganization of the railroad industry, the term *jinshin jiko* was not widely adopted until the early 1990s, when Tokyo began experiencing a marked increase in the number of commuter suicides. Prior to that, commuter suicides were typically reported as *tobikomi jisatsu* (suicide by jumping into the path of the train). The sudden rise in the number of commuter suicides was part of an overall trend in Japan beginning circa 1990 that saw the annual suicide rate consistently top 30,000.[5] Within the popular media and academic discourse, this sudden rise was regularly linked to the demoralizing effects of the collapse in the early 1990s of Japan's late-1980s bubble economy and the onset of a recession that dragged on for at least two decades.[6] Whatever the cause, the general effect of this increase in suicides was that in the late 1990s and throughout much of the first decade after the millennium, traffic on one line or another within Tokyo's commuter train network was daily brought to a halt by a commuter suicide. As one contributor to an online news forum remarked with unveiled exasperation in 2007, "Not a day passes without a report appearing in the newspaper of delay on one train line or another as a result of 'bodily accident.'"[7]

A commuter suicide creates an irremediable gap between the principal (*kihon*) and operational (*jisshi*) *daiya*, causing a crisis for the margin of indeterminacy. As a result, the train line is forced into a mode of recovery during which *daiya* technicians scramble to recalculate, redraw, and redistribute a new provisional *daiya* for each component of the system. Traffic cannot resume until this process is completed and the body has been removed from the tracks. Recovery is an exceedingly complicated process that draws on a *daiya* technician's intimate knowledge of the system, its rhythms, and its limits. Trains have to be cancelled, rerouted, and reordered in order to accommodate increasing platform crowds as quickly as possible, so as to prevent someone being inadvertently pushed to the tracks and yet another (human) body accident occurring. Depending on the time of day, the amount of time required for recovery varies. During off-peak hours, there is a good chance that the system can fully recover within a few hours. By contrast, a suicide during morning rush hours produces a disorder that spreads quickly throughout the system and resonates throughout the day. It is not uncommon to hear an announcement on multiple train lines late into the evening apologizing for a delay attributed to an incident earlier that morning.

It is tempting to read the introduction of the term *jinshin jiko* as part of an effort by the state in conjunction with the train companies to obfuscate the problem of commuter suicides. While not entirely improbable, such an interpretation is difficult to defend, since the designation "accident" in place of suicide actually aids surviving family members in collecting life insurance. It is hard to imagine that the state and train companies would conspire to incentivize commuter suicides by facilitating the life-insurance collection process. It is more productive to ask how the term *jinshin jiko* mediates and mitigates the potential disorder eventuating from the body on the tracks.

As an act that leads to the destruction of the actor, suicide makes for a difficult object of analysis. In treating suicide empirically, one can only attend to its representations, its effects, and its social and discursive determinants. Within the vast scholarship on suicide, the subject of suicide in Japan was for a long time treated in terms of culturally essentialist notions of uniquely Japanese social pressures and traditional mores, with authors sometimes referring to the persistence of a Japanese *Bushido* warrior ethic to explain the supposed Japanese proclivity for suicide.[8] More recent work has made great strides toward de-essentializing suicide in Japan, emphasizing instead the inseparability of the phenomenon from historically contingent sociocultural and economic circumstances and medical discourse.[9] Commuter train suicides can be explainable to a certain extent through this literature; the Japan legal scholar Mark West, for example, begins his chapter on debt suicide in Japan with a man contemplating jumping in front of an express train.[10] But to explain commuter train suicides through this lens is to dilute their ontological resonance within the symbolic currency of a general operation and thus lose the specificity of the phenomenon. This chapter takes up the matter of ontological resonance by thinking with the commuter train suicide in the context of the technicity of the commuter train network.

The technicity of the commuter train network is embodied in the *quality* of the collective engendered within the system's margin of indeterminacy; in this regard, it is an expression of the limits and possibilities of the co-constituting ontological entanglement of human and machine. Human and machine must remain spatially delineated for the collective to realize its operative metastability. Putting this in Simondon's language, we can say that the space between human and machine is a space between different orders of magnitude whose entangled functional coherence constitutes an associated milieu. At the edge of every train-station platform is a yellow line—or, more specifically, a yellow, tactile warning strip with bumps—that figures as a tangible instantiation of the spacing between the commuter train network and the commuter. The familiar refrain broadcast over platform speakers warns commuters time and again of the significance of this spacing: "*Momentarily, a Tokyo-bound rapid transit train will be entering the station. To avoid danger please stand back behind the yellow line.*" The potential "danger" of which the announcement warns is not just to the commuter but to the metastable integrity of the collective. Platform attendants and train drivers are thus forever tense, looking out for the wayward commuter during the morning rush who exploits the narrow yellow warning strip to slip past a long line, the commuter walking along while engrossed in their *keitai* or smartphone, or the drunk in the evening stumbling along at the threshold of the two disparate material orders. Such perilous actions prompt drivers to sound the horn as the train enters the station. But, as with the incident described above, drivers rarely have a chance to respond when a commuter decides to leap to their death in front of an oncoming train.

Japan's Ministry of Land, Infrastructure, Transport, and Tourism could not have chosen a more apposite term when it instituted the categorization of commuter train suicides as "[human] body accident." The commuter train suicide is indeed a disordering event generated by the human body on the tracks. Borrowing from Mary Douglas, we might think of the body on the tracks as "matter out of place" that threatens the integrity of the social order.[11] This would require, however, accepting Douglas's structuralist premise, which prioritizes a formally constituted symbolic order and equilibrium in order to then imagine their undoing through the boundary transgression. A commuter suicide is not a transgression of symbolically constituted boundaries. When the commuter body impacts with the commuter train, the critical spacing between human and machine is dissolved. The result is nothing short of a systemic crisis for the collective, as the disorder of the event propagates outward from the body on the tracks to collapse the margin of indeterminacy of the specific train line and disrupt the margins of indeterminacy for all intersecting train lines. By all accounts then, the body on the tracks should constitute an extreme event within the commuter train network and indicate a certain limit to its metastable viability. The paradox of the commuter suicide is that it fails to register as such. The forty to sixty minutes in which train traffic is stopped proves to be yet another regular irregularity, setting in motion a routine of recovery in the mode of repetition without difference.[12]

The commuter train suicide thus raises not only the question of how a commuter's death is normalized but also the question of how such a frequent and gruesome death becomes part of the regular order. Of course the easy answer to this question would be to say that the frequency of commuter train suicides has desensitized commuters to the eventfulness of the event. Yet frequency might also be expected to do just the opposite, making people question the ethical coherence of a system that seems to so easily accommodate death. How then is the disorder of the body on the tracks accommodated within the collective, and what kinds of collective becoming does its accommodation foreclose?

In chapter 3, I discussed JR East's introduction of novel, decentralized traffic-management technology that allows for accommodating

extreme operational events as regular irregularities. The new technology, I argue there, transforms the system into extreme infrastructure,
a system without limits that simultaneously enables the emergence of
extreme capitalism. The chapter focuses on the reconceptualization
and redesign of the commuter-train infrastructure in order to accommodate extreme operational events within the regular rhythm of operation. A similar concern guides my discussion in this chapter, although
here I specifically emphasize how the disorder engendered by the body
on the tracks impinges on the commuter train network's technicity.
It is important, in this regard, that the body on the tracks is not simply a technological or organizational challenge, such as a system malfunction or a surge in commuter demand would be. In collapsing the
space between human and machine, and with it the margin of indeterminacy, the body on the tracks presents an existential threat for the
collective integrity—a threat, more importantly, with ethical ramifications. For technicity does not concern merely the functional coherency
of collective. That is, it is irreducible to technological optimization,
understood in conventional terms as the hyper-rationalization of technical performance under capitalism's drive toward market efficiency.[13]
Technicity concerns rather a kind of ethical unity as it embodies the
quality of the collective's constitutive relationality and speaks to the
potential reticular emergence of mutual becomings.[14] When the irregularity of the body on the tracks is accommodated within the regular
order of operation, there are no becomings: a commuter is dead, and,
insofar as the operational performance of the collective is diminished,
no demand is made on commuters to confront that death as an ethical
dilemma. What would it take for it to be challenged as such?

My concern in this chapter is with the ethical challenge that the
body on the tracks poses to the technicity of the commuter train network. My approach is twofold: the first part of the chapter asks how
the limit embodied by the body on the tracks is overcome to produce
a functionally coherent yet ethically impaired collective condition.
In answering this question, I look at what I call the logic of recognition within the salaryman narrative of commuter suicide, in which the
notion that "it's always a salaryman" who commits suicide leaves the
body on the tracks unseen and unconsidered. The logic of recognition,
I argue, thus circumvents acknowledgment of the body on the tracks.
In so doing, it works to sustain a collective of integrated connections
without relationships, or what we might call a technicity without quality. Whereas recognition, I argue, is an act of determining a connec-

tion, acknowledgment demands the creation of an affordance for relationship by means of producing a gap of becoming within oneself. In
the second section of this chapter, I explore how the body on the tracks
stages a return, haunting the gap as a material force that disrupts the
mechanics of recognition and demands acknowledgment. My discussion there first parses the statements and writings of a former JR East
employee who was forced to (re)encounter the body on the tracks as
part of the crew dispatched to clean up after a commuter suicide. It
then moves to a close reading of the film *Suicide Circle*. The film, I argue, brings together this chapter's discussion both by foregrounding
the possibility of the eventfulness of commuter suicide and by making the matter of connection without relationship its central concern.
The latter is embedded in a critique of urban Japan (especially Tokyo)
and the nation's mass-media industry for its production of the problem
of group suicide as a media spectacle. Insofar as *Suicide Circle* voices a
strong critique of media, it is not anti-technology. It wants instead to
imagine a different kind of network, one founded on acknowledgment.

Repetition and Recovery

In forcing train traffic to a halt, a commuter suicide produces a stop-
page in the regular rhythmic tensions between structure and process,
control and emergence in the commuter train network. In so doing, it
sets into motion a different set of rhythms and tensions for commuters,
commuter-train operators, police, and firefighters toward the recon-
stitution of a provisional collective order amid the threat of systemic
unravelling.

In a single moment on that morning of August 19, 2004, the atmosphere of the collective shifted to a state of what could only be called
a routine crisis of disorder. The body on the tracks had halted all train
traffic on the Chūō Line between Tokyo and Tachikawa (a main commuter artery) at the peak of the morning rush hour, leaving 110,000
commuters instantly stranded.[15] The lucky ones were stranded at stations, which allowed them to look for alternative transportation via
parallel train lines and buses. The unlucky ones were stuck in packed
trains halted between stations. Upon hearing the announcement of a
jinshin jiko, commuters adjusted to the sudden change in circumstance
with a calm but intent proficiency. A large number of them began making their way back up the stairs from the platform to the main station.

As they exited the electric ticket gates, station attendants positioned on either side offered "Proof of Delay" slips (*chien shōmeisho*) to those who wanted them.

A "Proof of Delay" slip is a standard form distributed whenever there is a significant disorder of the *daiya*. It can be presented at a ticket office at any time for fare reimbursement and also serves as evidence that one's tardiness to work was not on account of one's own actions. Many of those who received the slip immediately entered a long line in front of the ticket office to receive reimbursement. Others joined another line stretching down the station stairway to the street to catch a taxi to a nearby station on a different train line. Even as commuters steadily exited the station, new commuters continued to arrive. The platform as well as areas around the ticket gates quickly became packed with slightly irritated-looking but quiet commuters.

As commuters and commuter-train operators adjusted to new circumstances, police and firefighters arrived simultaneously under a hail of sirens. The involvement of both is required according to Japan's Ministry of Land, Infrastructure, Transport, and Tourism. Both are part of the routine procedure of recovery. The police's task is to determine the nature of the accident, which two middle-aged officers began doing by talking first to the driver and then to eyewitnesses among waiting commuters. While none of the waiting commuters approached them to offer an account, neither did anyone refuse to talk when approached. From what I managed to overhear, the police were interested in the man's behavior before the event. Most likely they were trying to discern whether it was an accident or a suicide. Overall, there was no noticeable tension surrounding these exchanges, and the atmosphere on the platform felt, to me at least, uncannily flat.

While the police pursued their investigation, a group of firefighters, assisted by two station personnel, began the far more unpleasant task of removing the body from under the train. On Japan's *2Channel* textboard website, *jinshin jiko* is known as *gumo* or *geba*. The terms play on the phonetic and aesthetically disturbing quality of the words to convey the ghastly post-humanity of commuter suicide in a literal sense as an event involving body and machine that leaves the body no longer recognizably human. Among railroad employees, a common term for the body recovered from a suicide is *maguro* (tuna), which evokes the image of a deep red chunk of tuna meat split open in the market for display. The mangled post-human body, as we will see later in this chapter, is an intensive charge capable of interrupting the routine of recovery as repetition without difference. For the sake of repetition and recovery, it must never be seen. As a result, *jinshin jiko* is one of the most widely reported but never represented facets of daily life in Tokyo. While announcements of a *jinshin jiko* are often broadcast within the commuter train network throughout the day and reports of a *jinshin jiko* typically appear in newspapers the same day or a day later, that which is reported—the body on the tracks—is never seen.

In order to shield waiting commuters from the potentially gruesome scene and its disturbing affective resonances, firefighters or station employees commonly hold up a large blue plastic tarp while collecting the body from the tracks or from under the train. On that morning in Higashi-Koganei station, firefighters managed without the tarp by working from the opposite side of the train, out of the view of commuters standing on the platform. Although there was nothing to prevent someone from approaching the platform edge to get a better glimpse of the firefighters at work, no one did. Everyone stood queued at the designated areas, doing what they would have done had they been on the train: reading the newspaper, writing messages, playing games on their phones, or just staring off into space while plugged into headphones. Every couple of minutes, an announcement over the station public-address system provided an update.

At the time of the accident, Higashi-Koganei Station was being expanded as part of a decade-long reconstruction project to raise the Chūō Line and increase the number of tracks. As part of this project, a temporary platform on the opposite side of the tracks had been constructed but not yet opened for use. Taking advantage of the clear space there, firefighters laid out the body retrieved from the tracks and covered it with a white cloth. Only a small shock of black hair on one end and black dress shoes on the other were visible under the thin white blanket, which was stained in places with blood. A railroad employee later brought over a black briefcase that the man had left on the platform. Their work finished, the firefighters sat down a few meters away from the body. As they waited for the police, who were still busy recording details in small notepads, they passed the time in conversation, drinking cans of cold coffee from the vending machines. In the meantime, back on the platform, a gray-haired station worker arrived with two small buckets of water, which he splashed onto the front of the train in an attempt to remove a splotch of blood. The first two buckets of water had little effect, and the grey-haired man was forced to make three trips back to the station bathrooms in order to refill the buckets and repeat the process before the red smudges finally washed off.

Always a Salaryman

Train service resumed on the morning of August 19, 2004, at 8:26 a.m.—exactly forty-four minutes after the body on the tracks had brought traffic to a halt. The train involved was finally allowed to complete its entrance into the station, where it came to rest at the designated mark on the platform. If the commuters who had been trapped inside the train for almost three-quarters of an hour were relieved to be moving again, they did not show it when the train doors opened. Most of them remained on board, leaving little room for the crowd of commuters that had gathered on the platform in the time that traffic had been halted. Nevertheless, some commuters on the platform did manage to squeeze into the thick mass of bodies. As always, everything was performed without a single word spoken, aside from an occasional "excuse me," almost imperceptibly uttered. As soon as the train pulled out of the station, another pulled in, commencing a process of rapid arrivals and departures that went on for at least another half hour. Each train was as packed as the one before it and able to take on only a small number of new commuters from the platform. But little by little, the crowded platform began clearing.

During the forty-four minutes that traffic was halted, I moved among the gathering crowd of commuters on the platform in an attempt to engage whomever I could in conversation. I had no specific questions. I hoped only to glean an understanding of how commuters were experiencing and processing the incident. Finding someone willing to speak with me proved far more difficult than I had anticipated, as the incident seemed only to amplify the general disinclination toward conversation among the rush-hour commuters. For the most part, commuters waited patiently in line, reading books or magazines or focusing on their *keitai*. The scene felt decisively anticlimactic. Everyone seemed unperturbed despite knowing that a life had been extinguished in the most gruesome way. Although the suicide had enfolded us all in a profound disorder, there was no apparent atmosphere of event, only a subdued sense of impatience with the circumstances. This left me with no obvious means to start a conversation. My initial attempts at a straightforward approach ("Excuse me, I'm studying Tokyo's commuter train network and I was wondering what you think of what has happened") were met only with a quick wave of the hand to signal an unwillingness to talk. Less direct attempts in which I simply asked, "Does this happen a lot?" did not fare much better. A young woman told me,

"We're used to it," before turning back to her *keitai*, and a middle-aged man in a suit explained, "It happens a lot on the Chūō Line and it is extremely inconvenient. I wish they would stop."

After train service finally resumed, I noticed a man who looked to be in his late twenties leaning against the back wall of the platform reading a pocket-size novel. He was wearing a white button-down shirt and dress slacks, but no tie or jacket (typical for that hot season). As trains entered and stopped, he would look up from the novel just long enough to survey the crowd before returning to his reading. "I'm waiting for the trains to empty," explained the man, who introduced himself as Shimabukuro, when I inquired why he did not board with the other commuters. Perhaps because he was not in a hurry, nor was he seemingly irritated by being late to work at the accounting firm in Yotsuya where he told me he was employed, he was more willing than others I had approached that morning to engage in conversation. He explained that everyone in his office almost expected him to be late once or twice a week now because of the problems on the Chūō Line. "If it's not a *jinshin jiko*, then it's signal trouble or something else with this line," he added, albeit in a tone that seemed devoid of the kind of frustration one might expect under such circumstances. I asked why he thought someone would choose to jump in front of the train, especially at the height of the morning rush. He responded after taking only a few seconds to reflect on the question:

They do it during this time in order to elicit an appeal to their own existence, probably because they want to create annoyance. It's always a salaryman, and it's when they just snap, when they can't take it anymore after failing no matter how much they try and try. There was another accident not too long ago at this station, same thing, same situation, but the guy jumped from the far end in front of the special rapid. There might be more instances [of *jinshin jiko*] in the summer than in the winter, I'm not sure. At first, when I came to the city from Okinawa I thought that the way people just go on as if nothing happened was coldhearted, but then I realized that they are just used to it. Now the whole schedule is disordered for the rest of the day. And after all that careful calculation. No point in trying to check it from your *keitai*, maybe not even until they can restart tomorrow morning.[16]

For Shimabukuro, the body on the tracks was not a specific salaryman but rather the image of the salaryman in general. The latter arises in response to the question inherent in the first sentence: *What kind of person would need to elicit an appeal to their own existence?* It requires only a series of minimal mental gestures to form a compelling image connecting the body on the tracks with countless images in films and television dramas of the distressed salaryman caught in a downward spiral of increasing existential angst after failing at work, in his relationships with his colleagues, and as a father or husband. These connections exist in an established configuration ready to explain away the events of the morning and the ensuing disorder. They provide a ready-made template through which commuters can recognize the body on the tracks without having to acknowledge it. Via the image of the salaryman, the commuter train suicide thus becomes a missed encounter.

For Deleuze in *Difference and Repetition*, encounter elicits a mode of conceptual engagement antithetical to thinking that traces received frameworks of thought to render a world always already recognizable.[17] A recognizable world in Deleuze's writing is a world in which thinking has become a slave to an idea of thought as both rational and emancipatory but at the end of the day is nothing more than a feedback loop unable to escape structuring presuppositions. Encounter, by contrast, is a destabilizing moment when the logic of recognition fails, sending the individual into an existential freefall unsupported by the rational explanations of normative discourse. Encounter engages an object in the world on ontological terms as a contingent event and intensity that can only be grasped affectively. Encounter, in these terms, is predicated not on an unmediated engagement but rather on an engagement that is unmediated by existing structures of representation.

Deleuze's thinking can be parsed in terms of the distinction I am making between recognition and acknowledgment. Recognition traces relations as mere connections, while acknowledgment embodies an encounter that elicits relationship. What is important are the different vectors of mediation involved. To *recognize* involves an act of identification in accordance with a received framework of understanding. Its vector of engagement moves outward from the person to constitute an object. To recognize the death of a commuter in the crowd, for example, requires only identifying that there is disorder and mobilizing the image of the salaryman to make rational sense of it. To *acknowledge* implies a vector of force that moves from the world into the individual. It necessitates allowing oneself to be affected and possibly transformed through an encounter. Acknowledgement thus makes explicit the act of opening oneself up to an other or an outside. It demands creating a gap in oneself for becoming. *Encounter* thus carries the connotation of eventfulness, in that it engenders a departure from an existing structure of experience that makes way for a different, perhaps uncertain, future.

The notion that someone jumps in front of a train to "elicit an appeal to their own existence" involves an act of recognition, albeit one that invokes irony by reducing the body on the tracks to a received understanding of suicide as an expression aimed at amplifying one's individual autonomy and agency. Whether Shimabukuro meant to draw attention to the ironic outcome of such an appeal in the case of the commuter suicide is unclear. The irony is that, indeed, there is no more powerful means for commuters in Tokyo to assert their existence than to impact the lives of hundreds of thousands of commuters simultaneously. And yet, because of the logic of recognition that is mobilized, the event has exactly the opposite effect: commuters' attentions are drawn specifically to the inconvenience engendered by the act and not to the extinguished individual's existence. The irony at work here recalls Thomas Osborne's discussion of public suicide among literary and artistic figures as a unique expression of will.[18] Such public suicides, Osborne argues, demand to be interpreted as aesthetic performances through their scenography. Osborne turns to Mishima Yukio's famously elaborate and televised suicide in 1970 as an exemplary case for which "scenography was everything." Mishima's performance was a successful "aesthetic suicide," he argues, precisely because of its capacity to "impress a mark upon the course of time" and in so doing *"prolong* the presence of the will rather than erase it."[19]

Osborne's recourse to Mishima is supremely apposite to the discussion of *jinshin jiko* as a performance, although probably not in ways that Osborne would have intended. For scenography was certainly everything for Mishima. In fact, Mishima rehearsed his death many times in elaborately staged performances for the camera in order to refine the scenography for his final appearance. In the end, though, it was not exactly the scenography that defined Mishima's act, making it questionable whether he succeeded in prolonging the presence of his will. Mishima's intention was to generate encounter and event. Yumiko Iida writes of the performance: "In Mishima's own rationalization, the sensational suicide pact was an act of protest, an expression of rage at the spiritual degradation of postwar Japan, and a demand for the cultural revitalization of the Japanese."[20] This was the gist of a speech that Mishima delivered from the balcony of the building that he, along with four of his men, occupied by force on a Japanese Self-Defense Force base. Mishima sought to spark an event and propagate a disordering force capable of interrupting what he perceived to be the spiritually impoverished complacency of postwar capitalist Japan. He wanted nothing less than to change the course of Japan's history, putting the nation back on track toward right-wing militant nationalism. Ironically, the great author is not remembered for the final words that he delivered in a speech from the balcony of the building—those were drowned out by the pulsing rotors and whining engine of a Japanese media helicopter. Instead, Mishima is recognized for creating, in his last moments of life, a media event extraordinaire that fed easily into an emerging culture industry in Japan intent on manufacturing the affect of eventfulness for weekly consumption as a commodity spectacle within an interlocking network of television, magazines, and newspapers. To imagine that Mishima succeeded in prolonging his will requires a reduction of the complexity of conditions in which his drama was encompassed in order to attribute to him alone a certain agency.[21] His performance definitely left a mark on time, but not in the form of a prolonged will. What Mishima produced in the end was exactly the kind of phenomenon he had spent years decrying as a sign of the nation's spiritual collapse. His death enacted repetition without difference in a mass-media performance that fabricated connections without being able to demand acknowledgement. As we will see later in this chapter, it is this mode of media performance that is problematized in the film *Suicide Circle*.

Like Mishima's performance, the commuter suicide does not succeed as an appeal to an individual's existence because nobody seems to care about that. As Shimabukuro's comment suggests, the only mark it leaves on time is the disorder that propagates outward from the body on the tracks for much of the day. With the start of the next day, the *daiya* is reset and the disorder is forgotten. It has no lasting impact on the structure of the everyday, let alone on the unfolding of history. The body on the tracks, like Mishima's suicide, becomes a nonevent. What remains for the typical commuter stranded at some station to think about is only the actual functional transit connections between alternative stations, bus lines, and taxis in order to circumvent the disorder.

The Salaryman Narrative

"It's always a salaryman," Shimabukuro had said without hesitation. I decided to pursue the connection further. In the evening on the day after the *jinshin jiko* at Higashi-Koganei Station, I stopped in at small *yakitori ya* near the station. A *yakitori ya* is a tavern-like establishment specializing in reasonably priced grilled-meat skewers and other snacks and side dishes to accompany alcohol. Such establishments have long been associated with the salaryman lifestyle, and one is certain to find a number of *yakitori ya* around any train station in the city. Smaller *yakitori ya* around city-suburb stations like Higashi-Koganei tend to have a regular clientele. Some of them are young, unmarried men who do not want to cook and eat alone in their small apartments. Others are older, family men who live in the area and stop in occasionally for a quick snack and a drink before walking or riding their bicycles from the train station to their homes. These men tend to know each other's faces but do not necessarily talk to each other, reproducing, as it were, the conditions of the packed train. They sit alone at the counter reading a newspaper or book or just watching the obligatory television in the corner, which is usually tuned to the evening news, a variety show, or a drama. There might also be a few tables occupied by coworkers or university friends having drinks. I made it a point to become a regular at a *yakitori ya*, stopping by once a week always around the same time so as to get to know the other regulars. My *yakitori ya* was the simplest among four similar establishments around Higashi-Koganei Station, with only a large sign that read *yakitori* in Kanji characters over the entrance. It was a long, narrow space with something of a hole-in-the-wall feel but with all the *yakitori ya* essentials, including an old television in the corner and reasonably priced drinks. The owner was a slender, bald man in his late fifties named Takahashi. He always wore a clean white shirt and a blue apron around his waist. On my first visit, I had the chance to explain to Takahashi the nature of my research. Since then he had taken it upon himself to try to assist me by introducing me each week to the regulars.

Over a glass of beer and a plate of chicken-heart and liver skewers, I asked Takahashi if he knew anything about the suicide (*tobikomi jisatsu*) the other day at the station. Takahashi seemed confused for a moment by my use of the term *suicide*. "You mean the *jinshin*?" he asked, abbreviating the already abbreviated term to its bare minimum of "human body." Without waiting for a response, he went on: "That always happens on paydays: the fifth, fifteenth, and twenty-fifth of the month. That's when the money goes in but also when money goes out." "So *jinshin* are about money?" I suggested. Takahashi agreed, adding that the family of the jumper would be able to get life-insurance money. "Wouldn't they lose it all when the train company sues the family for its financial loss?" I asked, repeating information I had gleaned from a newspaper article. Takahashi corrected me: JR East did not make the family pay unless it received an unusually large sum from the insurance company.[22]

I asked Takahashi if he knew the guy who had jumped in front of the train the other day and whether he was in debt too. Takahashi used the question as an opportunity to draw the man sitting next to me into the conversation. He was a tired-looking, heavyset man nursing a *shōchu* cocktail and wearing a suit that looked like it needed to be cleaned and ironed. "That guy?" he responded, obviously having been listening to our conversation. "Yeah, I think he was a salaryman who borrowed money from one of those consumer finance outfits," he said, referring to a type of finance company that had emerged in the 1960s to encourage consumption among salarymen and quickly received the reputation of being a quasi-legal loan-shark operation.[23] That was all the information I could gather from either Takahashi or the heavyset man. Neither knew the name of the man who had jumped in front of the train, and neither could recall where they might have heard that he was a salaryman with financial problems. "It must take a lot of guts to jump in front of a train," I mused aloud, not sure what else to say. "You mean?" Takahashi joked, putting his palms together over his head and leaning slightly forward as if ready to dive. "That's not guts. It's something else."[24] Takahashi was called away by another customer looking for a refill on his beer before I had the chance to ask what he meant.

Half a year later I had the opportunity to confirm some of what Takahashi told me while interviewing JR East's public-relations representative at the company's central office in Shinjuku.[25] Although JR East publicly states that it seeks compensation for lost profit, it operates on a case-by-case basis. If, for example, an accident involves a vehicle from a company at a railroad crossing, then JR East bills the company for damages to the train and for lost profit during the time that service was suspended. If the accident involves a private citizen at a train station, then JR East might send out a bill only as a matter of protocol. The JR East representative emphasized that the company does not really expect payment, especially since few families have access to such enormous sums. He added that if the family is in severe debt, JR East might not even send the bill, simply out of consideration for the family's hardship.

A Stabilizing Narrative

Statistical data on *jinshin jiko* lists no specifics about the employment of those deceased; it states merely that the ratio of men to women is two to one.[26] Why, then, is it so easy to think that "It's always a salaryman" who commit suicide in the commuter train network? Insofar as the claim is of questionable empirical value, it proves to be a powerful meme able to contain the disordering force of *jinshin jiko*. Providing an effective image and motive behind the suicide, it stabilizes the potential disordering effect of the body on the tracks and marks it as a mere regular irregularity within a system without limits.

The figure of the demoralized, depressed, overworked, burned-out, and suicidal salaryman falls into what the medical anthropologist Junko Kitanaka identifies as a gendered discourse of depression in Japan.[27] This discourse, Kitanaka shows, informs the different ways in which men and women embody depression, the different kinds of recognition they receive, and the different kinds of access they have to treatment. The discourse does a considerable amount of social labor, producing men as workers and thus legitimate subjects of depression entitled to particular forms of state care and remediation. The depressed female worker, by contrast, is excluded from structures of institutional recognition. As a result, she must struggle for the right to be overworked and depressed before she can claim the right to receive care. Kitanaka suggests this may be changing, however, as a result of the increasing presence of global pharmaceutical companies in Japan and their tactical, pedagogical elicitations of female consumers.

With the onset of economic recession in Japan in the 1990s and the subsequent rise in the number of *jinshin jiko*, the figure of the unemployed or "restructured" (*ristora sareta*) and depressed middle-aged salaryman, rather than the overworked salaryman, dominated both academic and popular discourse. An article from a 1998 copy of *Aera Weekly Magazine* offering advice for overcoming midlife crises provides an example of this narrative in an account of four men who committed suicide by jumping in front of the train on Japan's Labor Thanksgiving Day (a national holiday commemorating labor) in November of that year.[28] All four men were in their early fifties and unemployed, the article relates, and none of them left a suicide note. Two of the men were also dressed in suits, it adds, despite the fact that they did not have steady employment and it was a national holiday. The absence of an unequivocal explanation for the suicides that a note might have provided becomes an invitation for the author to infer from the men's unemployed status and age that all four were restructured salarymen suffering from middle-age depression. Suicidal depression, the article warns, has become a serious threat for restructured salarymen, and the fact that two of the men were dressed in their work suits suggests that they clung until the end to an identity and routine that had once given them a sense of pride and security in life. This image of the restructured, suicidal salaryman clinging to the illusion of routine and normalcy embodied in his suit received international attention a decade later in Kurosawa Kiyoshi's film *Tokyo Sonata* (2008), in which the central male character is a laid-off salaryman who leaves his wife and kid each morning dressed in his business suit only to spend the day sitting in a city park among other restructured salarymen.[29]

When JR East decided in the early 2000s to attempt to mitigate the rising number of *jinshin jiko* through infrastructural measures, the image of the unemployed and depressed salaryman informed its tactics. One of the measures that the company took was to install extra-bright illumination at the ends of platforms to fill formerly dark spaces with light. "When people are in the dark they tend to brood over things and become pessimistic about life," explained JR East's public-relations representative.[30] Large mirrors were also installed on the wall opposite the platform. The rationale behind the mirrors, according to the JR East representative, was to pull the brooding salaryman back from the brink by forcing him to reflect, literally and metaphorically, as he stared across the seeming abyss.

There is no evidence that either of these measures had any effect in reducing the number of commuter suicides.[31] That is to be expected. Their efficacy lay elsewhere, as rhetorical props in service of a discourse that allows commuters to connect the body on the tracks with the idea that "It's always a salaryman." On the one hand, the conditions of possibility for this discourse were established long ago in the association of the packed train and the salaryman lifestyle. On the other hand, the notion that "It's always a salaryman" reflects another tendency in the national-cultural formation of Japan to produce the salaryman as a figure of iconic excess, embodying the nation's sociocultural pathologies. Each of the four circumstances of *jinshin jiko*—suicide (*tobikomi jisatsu*), entering the track area (*tachiiri*), falling from the platform (*tenraku*), and unintended contact with a train entering the station (*sesshoku*)—becomes recognizable through behavior associated with the excess of the salaryman lifestyle: the salaryman fell to the tracks or entered the track area because he was drunk (because salarymen always get drunk); the salaryman got hit while on the platform because the station was crowded during the morning rush with salarymen heading to work; and, finally, the salaryman committed suicide because he was unemployed, owed money to a loan shark, or was depressed. The source of this narrative is in part the Japanese media, which never seems to grow weary of aligning the salaryman with some new instance of aberrational behavior. In times of affluence, the salaryman is the pathetic lecher, shelling out ridiculous sums of money to date, and possibly have sex with, high-school girls. In times of economic recession and corporate downsizing, the salaryman becomes the victim of state policy that favors Japan's corporations and renders him an expendable worker-drone.

As the bearer of social excess, the salaryman recalls what the French twentieth-century scholar Georges Bataille called the "accursed share." The notion of the "accursed share"—a social surplus of energy that must not be redeployed toward profit but rather must be destroyed in extravagant ways—held an important place in Bataille's thought. The concept was part of Bataille's attempt to reframe a classic anthropological interpretation of sacrifice through the logic of thermodynamics in order to produce a critique of capitalism.[32] For Bataille, the lavish sacrifice of slaves procured in battle was a means by which premodern societies resolved their "accursed share" in accordance with the general economy of laws of nature and the cosmos. Such sacrifices were purely wasteful and thus an instantiation of an investment without return.[33] Modern capitalist society, for Bataille, is an abomination, a corruption of the cosmological order by virtue of its formulation of a "restrictive economy" dictating that all investments must generate a profitable return.

At first glance, there is perhaps no clearer image of extravagant and wasteful expenditure than the laboring and overly productive salaryman body consumed beneath the unforgiving wheels of the commuter train. And yet *jinshin jiko* is not performed as a sacrifice, because the salaryman is not imbued with the symbolic excessive value of the accused share. A proper sacrificial object is an ordinary thing that has been separated from the everyday order, purified, and treated with extravagance in order to render it sacred and unique prior to its destruction. The salaryman, by contrast, embodies the mundane and everyday. He is made expendable precisely because he is thought to harbor no unique value. He is a mere analog of the serialized commodity—mass produced and disposable. Moreover, Bataille's theorization of sacrifice within the logic of general economy was about maintaining limits in accordance with the functional equilibrium of a cosmological order. Sacrifice of the accursed share figures in this regard as a necessary diversion of energy from growth back into the "circuit of cosmic energy."[34] The salaryman narrative inverts this logic in the service of a restrictive economy. Containment of the disordering force propagating from the body on the tracks is about enabling a system without limits.

Whereas Bataille invoked the cosmological order to specify an economic limit, JR East attempted to summon a cosmological power for the exact opposite. On November 30, 1995, reports *Asahi* newspaper, a group of fifty JR East stationmasters from four different areas between Tokyo and Sagamiko Station on the Chūō Line gathered at different Shinto shrines along the train line to cleanse the system of its excess impurities by performing the Shinto ritual of *oharai*.[35] Submission to the ritual of *oharai* does not imply recognition of one's responsibility for those impurities; it tends to be just the opposite, in fact. The ritual of *oharai* is performed when the causes behind misfortune are indeterminate.[36] According to the article, the stationmaster from Tachikawa explained that having the ritual performed was something of a desperate measure to curb what appeared to be an unexplainable increase in incidents that year. The aim of the ritual was to unburden the limit imposed by the body on the tracks. It did not work. An article in the same paper a few weeks later reported that the invocation of ritual was completely ineffective.[37]

Recovering the Body on the Tracks

When Deleuze and Guattari channel Spinoza to suggest the body's importance lies not in what it is but in what it can do, they shift the focus from the body as a site of identity and essence to the body as something that can affect and be affected.[38] Affect, in this regard, underscores the body as a coherent set of processes—a kind of machine—with the capacity to effect change and itself be changed in encounters with other bodies, whether human or nonhuman. But as the former JR East train mechanic Satō Mitsuru explained to me, the body on the tracks is often something of a non-body, no longer recognizable as human. Nevertheless, it retains a powerful capacity to affect. In Satō's accounts of washing and inspecting trains involved in commuter suicides, the pulverized body's affective capacity emerges to force an encounter that demands acknowledgment. Such moments figure as the return of the body per se in ways that foreshadow my discussion of *Suicide Circle* in the final part of this chapter.

Satō worked as a JR East mechanic for four years. For most of this time, he was stationed in a train garage in Chiba, a prefecture bordering Tokyo proper and home to a large number of commuters. After quitting the position, Satō began publishing accounts of his experience at the train garage, first in a blog titled *Behind the Scenes of the Railroad*

World (*Tetsudō gyōkai no butai ura*), and later in book form.[39] I met with Satō a few months after his first book went into its second printing, on a Saturday afternoon in Tokyo in September 2012, as a typhoon was bearing down on the city, drenching the streets with rain. We met near Shinjuku Station's east exit, at one of the chain-restaurant cafés that cater to middle-aged men and tend to be inundated with cigarette smoke. Satō appeared younger than I had expected. He was slender, with thick black hair, and wore a clean white button-down shirt tucked neatly into black slacks. We talked for nearly two hours, during which Satō elaborated on the experiences he had written about on his blog and in his books.

Satō's writings regarding his experience with JR East range from detailed accounts of the more trivial aspects of commuter-train operation to critical reflections on JR East. It makes for fascinating reading, not only for the wealth of information he provides on JR East, but also for an implicit social concern that inflects his descriptions. Satō's description of the cleanup and inspection after a commuter suicide is particularly interesting for the tension it conveys in the necessity for the workers to recognize the body on the tracks as part of the mundane regular irregularity of their work while avoiding an encounter that might force acknowledgment. Recognition, in this regard, functions as a coping mechanism. For Satō, that mechanism rests on his capacity to maintain the presupposition that the body on the tracks was a salaryman. When faced with the appearance of physical features incommensurable with this presupposition, that capacity fails and Satō is cast onto the uncertain ground of encounter. At such points his detached description of the gory scene falters and is punctuated with expressions of shock, concern, and discomfort that invoke both pity and ritual practices of purification. At such moments the affective capacity of the body on the tracks emerges and *jinshin jiko* begins to register as event.

The "*Maguro* Train"

Satō's first experience cleaning up after a commuter suicide occurred during his second year at the train garage in Chiba. Arriving one morning to work he found his coworkers anxiously awaiting the imminent arrival of a "*maguro* train," which he explains is the mechanics' idiom of choice for a train that has been involved in a commuter suicide and taken out of service to be cleaned and checked for damage. As usual, the suicide had brought train traffic to a halt on the train line but the rumor going around among the workers was that it was a particularly bad *maguro* train. Very often, Satō explained, the body bounces off a hard-plastic protective skirt covering the wheels on the front of the train. Such cases are called "skirts," and the damage tends to be minimal, with the body more or less in one piece. Particularly bad cases of commuter suicide are when the body somehow makes it under the skirt and gets wrapped up in the wheels and axles—what Satō and the others were expecting that day. In his written account of the incident, Satō describes himself and the other mechanics waiting anxiously for the *maguro* train while looking on silently as the *daiya* technicians, hunched over a printout of the *daiya* spread out on the floor, struggled to recalculate a recovery from the disorder.

Satō takes a cool and composed tone in his written account. "Commuter suicides are just part and parcel of a railroad worker's experience," he writes. "Until you've dealt with one your experience as a railroad worker is incomplete." Describing the work of cleaning up the body as "unsettling" yet also somewhat morbidly fascinating, he contends that "one doesn't get any feeling of lament for the deceased since commuter suicide is just so much a part of regular railroad work."[40] Yet in an action that seemingly contradicts that sentiment, Satō follows the advice of one of the veteran workers and grabs a bag of cooking salt from the kitchen as he and the other workers are departing the office to meet the arriving *maguro* train. Salt is used for ritual purification in Japan, especially in Shinto-related practices or ceremonies involving the dead. When returning from a funeral or gravesite, it is common for individuals to scatter salt over themselves or place mounds of salt around the house. Spreading salt delineates boundaries between the living and the dead. It is thus about recognizing the dead but is also a gesture that lends comfort by marking and containing an event that is specifically not part of the regular order of the everyday. Its efficacy on the scene of the *maguro* train, however, becomes uncertain as what is at stake are not boundaries but a process of human and machine entanglement that has collapsed in the fusion of flesh and metal.

On the site of the *maguro* train, Satō describes a grisly scene of pulverized and minced flesh, human fat glistening in the sunlight, and a horrible smell. Overwhelmed, he begins hastily tossing salt in every direction. "It felt like chanting a sutra and somehow made it easier to do the work," he explains.[41] That comfort, however, is short lived. It is interrupted when one of Satō's fellow workers points to long black hair mixed in with the pieces of flesh. The strands of long black hair force a moment of encounter, shattering the connections Satō has constructed to recognize the body on the tracks. Satō is taken aback and suddenly made aware that he had assumed the mangled body belonged to a tired and broken middle-aged salaryman. The long black hair renders that assumption unsustainable, and the new possibility that the body on the tracks was a young woman, someone maybe even around his age, forces its way into his mind. Having found an opening in the logic of recognition, the thought takes on life. Satō finds himself feeling somehow closer to the woman, feeling pity for her. A stream of questions that are impossible to answer floods his thoughts: Why did she do it? Did she not realize what collision with the train would do to her body? Could she not have imagined that? Or perhaps she did imagine it. What kind of person was she and who did she leave behind? If spreading salt marked his explicit recognition of the dead, the encounter with the long black hair threatens to demand acknowledgment of the unique life extinguished beneath the wheels of the train. The reader is left to wonder, however, what actions, reflections, and expressions of moral obligation this acknowledgment would entail as Satō does not allow himself to pursue the questions any further in his writing. His description of the encounter with the long black hair ends with him forcing his mind back to focus on the work as he returns to washing the train down with water from a hose and scrubbing it with a deck brush. The acknowledgment elicited from the long black hair was only temporary, it seems. It lacked the necessary intensity to issue an ethical challenge to the technicity of a collective in which the *maguro* train has been rendered a regular irregularity. Nevertheless, the body on the tracks is not entirely finished with Satō.

When the cleanup work is done and the tools returned to their places the crew takes a tea break. The atmosphere among the men takes on a lightheartedness, writes Satō, as they begin recalling moments from the work. For Satō, the respite proves once again only momentary. Asked to return to the train in order to check the brakes, he is horrified to discover a large group of crows congregated around the vehicle, carrying bits of human flesh in their beaks. Feeling suddenly nauseated at the sight of "ghastly black creatures consuming human flesh," the image of the long black hair returns, again forcing to consciousness questions that cannot be answered.[42] "That image of the crows is stuck with me even though I quit JR years ago," Satō confided over coffee in Shinjuku. "Even today I can't look at a crow without thinking about it."[43]

Satō continues to struggle for the rest of the day after the encounter with the *maguro* train with uncanny returns of images of the cleanup scene. When shopping for food at the supermarket after work he is anxious at first, he writes, wondering if he would be able to even eat. But upon discovering that he feels no "discomfort" when looking at the *maguro* tuna and raw slices of beef laid out in the display cases, he declares with pride in the next sentence, "it turns out that I'm okay with this kind of work after all." Later while preparing dinner, however, his confidence unravels when he spills minced pork on the flame. The sight of scattered minced meat and the smell of it burning in the fire instantly transports him back to the scene of the cleanup, sending chills throughout his body. Overcome with nausea he is about to throw everything in the garbage but suppresses the feeling and continues cooking. The encounter from the day is thus brought to a close through a forced resumption of repetition. "In the end, I made myself eat everything and it was really excellent," Satō declares, seemingly relieved, in the final lines of his account.[44]

Satō took pride in his work as a mechanic. He quit after only a few years, he told me, because the work stopped being challenging at some point. Even *maguro* trains, he said, became just part of the routine.

Re-mediating the Body on the Tracks

The 2002 Japanese horror/mystery film *Suicide Circle* (*Jisatsu sākuru*) foregrounds the problem of recognizing connections versus acknowledging relationships within a sharp critique of the quality of social life in contemporary urban Japan. Although *Suicide Circle* concerns suicide in Japan in general, the film's opening scene, in which fifty-four high-school girls leap to their deaths in front of an arriving train at Shinjuku Station, provides its formative spectacle and allegory; the film also revisits the station platform throughout as both an actual and figurative place of event. In addition, while we hear about group suicides in other parts of Japan at various points, the film never leaves Tokyo. It is thus in many ways a film specifically about Tokyo and the kinds of relationships afforded within its hyper-mediated and fast-moving culture of fashion and consumer trends. In this context, *Suicide Circle* elevates suicide in the commuter train network from a regular irregularity within everyday order to an event that encapsulates the pathologies of Tokyo's hyper-mediated environment, while also gesturing toward the potential emergence of a new and absolutely unrecognizable techno-social reality.

Suicide Circle's mass commuter-suicide scene unfolds in the opening title sequence, accompanied by a sorrowful tune. As the camera tacks back and forth between an approaching Tokyo-bound Chūō rapid train and Shinjuku Station on a typical weekday evening, we see a group of high-school girls descending the station stairs to platform eight. The sorrowful opening tune ends, and an intertitle announces the date, May 26. Dressed in the standard high-school uniform of white shirt and blue skirt, the girls occupy the platform in groups of two and three and are engaged in exuberant conversation. The camera then shifts to a jerky handheld perspective that lends the scene a documentary-like quality. The image becomes blurred at points as it moves among the groups with only sporadic words discernible in the background noise of youthful chatter. It is the kind of scene one sees daily on almost any platform, with absolutely nothing peculiar about it. However, the mood instantly changes following the platform chime and the announced arrival of the Tokyo-bound Chūō rapid train. The girls suddenly move, as if programmed to respond in unison, to the edge of the platform, where they simultaneously take one step in front of the yellow line before taking hold of each other's hands.

As the soundtrack switches to a light dance melody, the camera cuts back to the approaching train before panning across the human chain of high-school girls, their faces suddenly blank, their eyes gazing straight ahead. Swinging their arms together with a loud and vigorous cheer—"one, two, three!"—the girls leap all at once before the oncoming Chūō rapid train as it enters the station. The ensuing carnage is conveyed in the embellished slasher/horror film style of split-second

FIGURE 5.1. Fifty-four girls lining up on the platform to jump.
Source: *Jisatsu sākuru*. Directed by Sono Sion. Tokyo: Kadokawa-Daiei Pictures, 2002.

close-ups on body parts being crushed by the train, screams, gruesome sound effects, and a prodigious amount of blood that bursts through train windows and covers the station crowd and platform. *Suicide Circle* shows us what the television news and newspapers in Japan never will: the body on the tracks. Whereas the body on the tracks within normalized commuter suicide is recognized but never acknowledged, the bodies that have been pulverized and dismembered in their collision with the train become in the film an uncanny material referent for the emergence of a novel, technologically mediated relationality able to overcome the boundaries of time, space, and even death. We are given hints of this when, at the end of the opening mass-suicide scene, an unidentifiable figure leaves a white sports bag on the platform containing ten-centimeter strips of skin from the dead high-school girls and others that have been stitched carefully into one long roll. Similarly, we also learn later that the event has melded the girls' limbs and torsos into an inseparable and unrecognizable mass.

Written and directed by Sono Sion, *Suicide Circle* has become something of a cult classic, in part due to its unforgettable first scene. But its success also reflects the critique, mobilized by that scene, of the Japanese media industry's treatment of the nation's rising suicide rate. At the time that the film debuted, reports of "group suicide" facilitated through internet BBS and chat sites appeared regularly in newspapers and magazines in Japan, alongside articles by social theorists, critics, and health experts expressing alarm over an unprecedented rise in the nation's annual suicide rate above 30,000. The authors of these articles and figures in the mainstream media were particularly disturbed that participants in group suicides were increasingly young juveniles and minors who found one another via websites. On one level, the young age of participants resonated as a shocking rejection of national ideology. Amid constant media discussion of Japan's uncertain future as a result of the declining birth rate, the nation's children were choosing to die rather than accept the mantle to carry the nation and its culture into the future. Japan's few children were refusing, it seemed, the path mapped out for them by a generation of aging postwar leaders who had brought Japan from bubble to bust and increasingly appeared incompetent, corrupt, and incapable of dealing with the emerging complex reality of the information-networked world. On another level, group suicide suggested something at work in the relationship between Japan's younger generation and its digital machines that terrified the older generation. Whether in Japan or elsewhere in the world, one need not look far to find a technology discourse expressing anxiety over the anticipated loss of authentic human communication as a result of the younger generation's proclivity for digitally mediated interaction, whether via websites or phones.[45] Such discourse places an inherent ethical value on the supposedly unmediated quality of face-to-face communication while decrying the fraudulent nature of technologically mediated interaction. If the story of *Densha otoko* (chapter 4) gained traction for countering this discourse in its gesture toward the positive qualities of online communication, group suicide conveyed the opposite, suggesting not only something sinister about online communication but, moreover, that its sinister quality was specifically something that the older generation would never grasp and that it offered something deeper and more affective for its participants than face-to-face communications. Group suicide spoke to the terrifying possibility that technologically mediated connectivity lends itself to the formation of a new order of relationships among the young and their machines. In the film, suicide in the commuter train network becomes an exemplary act for the realization of that possibility as well as a means of critiquing the existing shallowness of Tokyo's massmediated society.

Not surprisingly, the group-suicide phenomenon was heavily discoursed in the Japanese mass media as one of the many "social problems" (*shakai mondai*) that arose during the post-bubble recession era otherwise known as "Japan's lost [two] decades." Other notable social problems that the media labeled and fueled during that era included bullying (*ijime mondai*), retirement funding for postwar baby boomers (*nenkin mondai*), and classroom collapse (*gakkyu hōkai mondai*). One could fill a book with just the designations and definitions of these various social problems. The discussion of such social problems tends to be the focus of daily news broadcasts and weekly news magazines. While the Japanese mass media is certainly not alone in its proclivity for generating sensationalistic news topics, the social problem is a particular sort of media object-issue produced within Japan's intensely orchestrated mass-media structure.

Comprising tightly affiliated companies, Japan's mass media industry produces a wide range of synchronized content across print, broadcast, and digital media. As a result, media channels tend to reiterate and reinforce the same message, creating what can feel like an intense feedback loop. While there is nothing Tokyo-specific about this structure of media, Tokyo's infrastructure environment becomes an intensive media ecology in service of these various channels. Advertisements for the various print and broadcast media invariably appear throughout

Tokyo's commuter trains and are refreshed on a weekly basis to keep all channels synchronized. The overall effect is a sense of total immersion within an intense collective interest around each social problem. One gets the feeling that all of society is deeply connected and that social problems are the focus of a genuine collective reflection. One could be forgiven for mistaking this apparent discussion around the weekly social problem as a kind of total social fact worthy of sustained ethnographic inquiry. Open any magazine or newspaper, turn on any news special, or glance at the table-of-contents advertisements for magazines hanging in trains, and everyone seems to be talking about the same social problem. All of society appears not only intimately connected and attuned, but also committed to genuine redress of the "problem." Despite this intensity, there is hardly ever a resolution to any single social problem. Instead, the social problems tend to appear and disappear on a weekly, sometimes monthly, basis, like fashion trends. Circulating on the same channels as weekly consumer fads, the social problem is thus a product of the same tightly coordinated mass-media structure that proved remarkably efficient in marketing commodities and setting trends throughout the postwar period and into the present.

Similar to the social problem and moving within the same media circuit is the "crime event" (*hanzai jiken*). Daily newspaper headlines and the table of contents of weekly and monthly magazines are also typically packed with news of various crime events, which might be violent crimes or scandals under investigation as possible crimes. Like the social problem, the crime event seems to demand attention and investigation. It is a happening that becomes a topic of conversation and concern as something that harbors a potential risk to the greater collective order. What is more, the term implies a certain level of organization, planning, or orchestration behind an event. As such, it alludes to the possibility that the crime event is part of some larger force or trend.

Suicide Circle takes up the issue of group suicide within a powerful critique of mass media's production of the social problem and crime event. In the film, the media-industry treatment of group suicide in these terms comes to reflect precisely what is wrong with Tokyo. It shows a society in which the mass media's capacity to drive the ceaseless recognition of one social-problem trend and crime event after another has created hyperconnectivity without relationships. The result is a society devoid of genuine and ethical relationality, in which the pain and emotions of others are recognized but never acknowledged.

Suicide Circle resists, however, falling back on a simplistic argument for the supposed authenticity of unmediated relationships. In the end, it asks us to think about the possibility of technological mediation driven by indexical relation to corporeal bonds. Bodies and machines become entangled in a radically new matrix of possibilities, for which we can only imagine, at this point, a trajectory of emergence.

Cause and Effect

Suicide Circle is not an easy film to watch. The difficulty it presents for the viewer has less to do with its macabre opening scene of mass suicide than its attempt to formulate its message in a bewilderingly complex series of loosely intersecting plots. Each plotline, moreover, contains its own specific trajectory of critique. In this context, the film presents at least two qualitatively different kinds of group suicide. The first are copycat group suicides, such as when several high-school students excited by the event in Shinjuku decide to commit suicide by jumping together off the roof of the school. Copycat suicides figure in the film as meaningless and pathetic acts. By contrast, the mass suicide at Shinjuku represents an entirely different kind of group suicide and event. Unlike the media-driven copycat suicide, it is not the effect of something but rather the cause of an emerging order that involves a suicide website (www.maru.ne.jp), a pop-idol girl group called "Desert," and a group of children that move in the girl group's orbit.[46] We are provided a hint of the complex relationship between these disparate forces when immediately following the event on the Shinjuku platform the camera pauses on the blood-soaked yellow tactile warning strip before shifting to the suicide website on which the image of the yellow warning strip is repeated in fifty-four yellow circles that start turning red amid enthusiastic clapping. The clapping, we soon realize, emanates from a television broadcast of a performance by the pop-idol girl group Desert. Wearing jersey dresses with different numbers stitched on them (the significance of which we learn about later), the five-member band of twelve-year-old girls performs complicated choreography, with the exaggerated enthusiasm typical of Japan's tightly structured media performances, as they sing a catchy tune that foreshadows the film's problematization of connectivity without relationality:

Mail me, hurry and hit the send key.
Can't you see I'm always waiting?

Mail me, wherever I am, I am always thinking about you.
Mail me, messages from others make me happy, too, but
Mail me, yours are special.
You probably don't know how I feel.[47]

On the one hand, the concert is a sharp parody of the kind of girl-band performances that are commonplace television spectacles in Japan. In the context of the opening sequence beginning with the mass suicide of young girls at Shinjuku, the cut to the concert scene resounds as a caustic critique of the media industry's tacit violence in commoditizing and eroticizing increasingly young female bodies. On the other hand, there is also something extraordinarily creepy about Desert's performance. With its catchy tune and explicit plea in an imperative form for connection, the song insinuates the work of a nefarious system of media contagion. Through an implicit intertextual reference to countless media-horror films, the viewer is left with the ominous feeling that the performance is some kind of novel media virus that exploits the mundane convergence of internet and television to transform schoolgirls into automatons ready to leap to their deaths with the reception of a single command sent to their *keitai*. The suspicion that the girl band is propagating a coded message of suicide via *keitai* will indeed turn out to be correct, although not in the way the viewer is initially led to imagine. Desert is not simply propagating connections to induce mass suicides; it is co-opting the nation's mundane channels of broadcast, communication, and transportation infrastructure to promote a novel kind of emergent relationality.

Intentionally misleading the viewer in this regard produces suspense. At the same time, this plot device operates on a more complex level to implicate the viewer in the media-industry structure the film sets out to critique. That is, in leading the viewer to assume a criminal connection between Desert's performance and the group suicide, the viewer comes to expect that the film will track down the connections and solve the riddle. Indeed, this desire to have the connections made and the puzzle solved leads the viewer to keep watching. But it is also what fuels popular interest in the weekly spectacle of scandalous crime events in the media. Being able to make connections merely mobilizes a structure of recognition operating within a given typology of actors and causes. As with the configuration of ready-made connections through which commuter suicides are recognized as a salaryman phenomenon, recognizing connections in the weekly scandal exposés demands no actual acknowledgment; no demand is made to accept the pain of an other through an encounter unburdened by received frameworks of thought. In sum, to expect that the film will solve the group-suicide puzzle is to reduce the phenomenon to a mystery commensurate with the banal weekly scandal propagated in newspapers and magazines and thus to become complicit in the exploitative structure of Japan's mass-media industry. *Suicide Circle* wants group suicide to be understood instead as the incipience of historical eventfulness marking social transformation.

This complex layering of critique makes *Suicide Circle* an extraordinary film. Like the coded message propagated by Desert, however, the film's message is obscured in its central plotline, which unfolds in the mode of a detective mystery around two characters—an older police detective named Kuroda and a high-school girl named Mitsuko—looking into the mass suicides. Kuroda's and Mitsuko's search for answers concerning the suicides takes them in a similar direction, putting them in contact with each other and with the group of children around Desert and the suicide website. Kuroda is motivated by a desire to identify the connections behind what he believes is an orchestrated crime. He thus embodies the media-industry norm, performing more or less in accordance with what the viewer expects in regard to the weekly scandal exposé. By contrast, Mitsuko is not merely looking to make connections and solve the riddle. She becomes involved in the film's events when her boyfriend commits suicide and is motivated by a desire to understand the suicides and her possible relationship with them.

Kuroda's and Mitsuko's differing methods and motivations produce drastically different results. For Kuroda, the search ends in catastrophe for himself and his family, while for Mitsuko it ends opening toward an uncertain but also potentially hopeful future. Kuroda, moreover, is a tragic figure in that he fails to understand in time a hint regarding the true nature of the group suicides that he is given in a telephone call from a boy who apparently suffers from a kind of Tourette's syndrome that compels him to constantly clear his throat. The boy tells Kuroda there will be another mass event that evening on the same platform and provides a clue, telling him to look for the sixth link in the chain, by which he means the roll of stitched-together skin left on the platform when the fifty-four girls jumped onto the tracks. The sixth piece of skin, it turns out, has a tattoo that none of the police, including Kuroda, can recognize. Unable to understand the clue, Kuroda and the other police detectives respond predictably, staking out the Shinjuku Chūō Line platform. Nothing happens, of course. Being too literal in their interpretation, the police fail to understand that "train platform" is a metaphor for the space and time of intense network connectivity that is the condition of possibility for an eventfulness of a different kind. In the meantime, as the police detectives are busy watching the Shinjuku Station platform, a simultaneous television and radio broadcast of Desert transforms all of the city into the platform. As a group of young people holds up signs in a busy Tokyo intersection, declaring "jump here," we see men, women, and children all over the city take their own lives in gruesome acts, accompanied by a saccharine and cheerful song from the Desert performance.

Only when Kuroda returns home later that evening and finds his dead children, including his son's body with a piece of skin missing from a tattoo that matches the tattoo from the sixth link of the chain, does he understand the significance of what has happened. Kuroda is devastated. He has not merely failed to recognize a clue; rather, he realizes that he has lived his life connected to but not in relationship with his family. As Kuroda sits slouched and broken against a wall on which is written, in giant characters, "This is the platform, jump here!" he receives a telephone call from the same boy as before. The telephone call drives home the film's underlying theme of recognition versus acknowledgment.

"We did not do it," the boy tells Kuroda before launching into a long series of rhetorical questions concerning relationships and empathy. Kuroda, unable to speak, just sits and listens with the phone to his ear:

What is the relationship between you and yourself?

Do you understand? I understand the relationship between you and me.

I understand the relationship between you and your wife.

I understand the relationship between you and your children.

As for the relationship between you and yourself . . .

If you die now, will you maintain the relationship to yourself?

Even if you die now, the relationship between you and your wife will not disappear.

Neither will your relationship with your children.

But if you die now, will the relationship between you and yourself disappear?

Will you survive? Are you a concerned party to yourself? [*Anata wa, anata no kankei-sha desuka*][48]

The boy begins by claiming a position of relationality and understanding that Kuroda has been unable to occupy: "Do you understand? I understand the relationship between you and me." While the subsequent enigmatic line of questioning is difficult to parse, it suggests a kind of relationality that exceeds recognized boundaries of time, space, and self. The significance of such relationality is made clear when another child comes on the line to deliver to Kuroda a stinging rebuke for his selfish inability to acknowledge the pain of others:

Why couldn't you feel the pain of others as you would your own?

Why couldn't you bear the concerns of others as you would your own?

You are a criminal.

You are scum who could only think for yourself!

Scum! [*Gesu yarō!*][49]

Kuroda's crime was pursuing the suicides without seizing the opportunity to examine his relationships with others and, in so doing, opening himself to them to acknowledge their pain. But he has realized this too late. Kuroda drops the phone and grabs a gun from one of his police colleagues. "It's no use. They [the kids] are not the enemy [*yatsura teki jya nai*]," he declares, putting the gun in his mouth and pulling the trigger.

In declaring that "the children" are not the enemy, Kuroda makes a generational distinction that forms the core of the film's critique and its message of hope. The children's statements condemn not only Kuroda but also his generation for the society they have built in which the deaths, misfortunes, and pain of others are turned into regular irregularities in the form of weekly spectacles devoid of any true social reflection or historical eventfulness. The children are not the enemy; the enemy is Japan's mass-media industry, which has produced a society of hyperconnected but morally and emotionally detached individuals. This critique proposes not a rejection of technological media but rather a rejection of a current technological order that shuns the disordering indeterminacy of extreme events for the sake of routine and profit, as Mitsuko's story makes clear.

Mitsuko's search into the suicides begins with her dead boyfriend, who, it turns out, was a Desert fan. While looking through his room, which is loaded with Desert paraphernalia, Mitsuko notices a poster of the Desert girls in their numbered jerseys. In a moment of uncanny

FIGURE 5.2. Mitsuko notices the poster.
Source: *Suicide Circle*. Directed by Sono Sion. Tokyo: Kadokawa-Daiei Pictures, 2002.

connectivity that suggests the children have unusual ability to communicate with and through various forms of networked communication technology, Mitsuko's boyfriend's landline phone begins ringing, together with his *keitai* (its ringtone is, of course, the song "Mail Me"), and following that the computer suddenly turns on. Mitsuko does not answer the phone, but the message seems to get through nevertheless, as she suddenly realizes that each band member in the poster holds up a different number of fingers, which have been circled in black pen, presumably by her boyfriend.

Mitsuko converts the combination of fingers and jersey numbers into keystrokes on a *keitai*. The word that is spelled out is "suicide." The moment the word appears on the *keitai* screen, the landline phone begins ringing again. This time Mitsuko picks it up, and a recorded voice asks her to input a password number. Mitsuko inputs the numbers from the girls' jerseys, and the boy who talked to Kuroda earlier comes on the line. "Do you understand," he says, "there's no suicide club. We haven't made such a thing. Come over and play with us," he says before hanging up. Mitsuko seems to understand immediately what she needs to do and begins writing down a string of numbers that appears on the computer screen. Her writing is interrupted (has she decoded another message?) by a message on the screen announcing a live Desert performance the next day.

Sneaking in backstage before the performance, Mitsuko encounters groups of children who subject her to an enigmatic line of questioning similar to that which Kuroda experienced. The children inquire, "Why did you come? Did you come to restore your relationship with yourself? Or did you come to sever that relationship? Are you a person without a relationship to yourself?" When Mitsuko insists, "I'm me and I'm in a relationship to myself," the children applaud her before issuing a series of statements that again suggests a kind of global and mutually constituting relationality irreducible to simple cause and effect: "Can you be in a relationship? A relationship between you and me, a relationship as between a victim and assailant? Can you be in a relationship as between yourself and a lover? Are you acknowledged by your own self?" "When rain dries clouds form, when clouds moisten rain forms," all the children repeat together.

Apparently satisfied with Mitsuko's answers, the children lead her to a dimly lit room filled with chirping baby chicks. Mitsuko, along with three other young women in various states of undress, kneel facing the wall as a masked man prepares a traditional wood planer and hands it to the children. The children then remove a strip of skin from each woman's back; each strip is then stitched into yet another thick roll. When, in the next scene, that roll of skin is delivered to the police via another suicide, we are prepared to see Mitsuko meet her end. Indeed, this is what we expect in the final scene as Mitsuko's and other high-school girls' *keitai* spontaneously ring in unison as the girls descend the stairs to the Chūō Line platform at Shinjuku station. But there is no mass suicide this time. Contrary to expectations, Mitsuko and the other girls board the train without incident as a concerned and perplexed police detective who was working with Kuroda looks on.

Something has happened, *Suicide Circle* seems to want to tell us: A message, it seems, has been delivered and an event has taken place. Something has changed or is changing, although the nature of the change remains unclear. What is clear is that the children have somehow forged a new kind of relationship with machines and with that a possibility for a relationality that exceeds the mere hyperconnectivity of Tokyo's mass-media industry. The children have taken control of the infrastructure and its technologies of communication. The city is theirs to transform into a "platform," allowing them to put into motion an alternative future that neither reveres nor needs the older generation. The film thus ends on a note of hope, with a last message from the (disbanding) girl-band Desert—"Live as you please!"—followed by a final televised performance and a farewell song that brings forward the idea of an emotionally charged relationality through and above technological connectivity:

While we idle away, not paying attention, we continue to press numerous command keys.
When you open your mouth, and reveal your feelings to me
What am I to do?
Yes, it's terrifying, but also so much fun too.
But we only have one chance, to light up the fire in each and every one of us.

Life is about lighting up once.
Romance is also about lighting up once.
Memories, too, are lighting up once.
So, we need to be a little more brave.

Once more, we need to start all over.
Sometimes I curse these disrupted feelings.
If you could love me with all of my emotional wrinkles.
If you could trace their grooves with me.

It's scary, true, but endearing.
This is a real goodbye.
I'll miss you.

Life is about lighting up once.
Romance is also about lighting up once.
Memories, too, are lighting up once.
But I want to forget about all of it.

Connections

"No act is more ambiguous than suicide, a riddle cast in the teeth of those who live on," wrote the French Japanologist Maurice Pinguet.[50] Commuter suicide in Tokyo's commuter train network is less a riddle than a statement demanding we focus our attention on the ethical integrity of a technicity that can process the body on the tracks as a regular irregularity. What emerges is a technical ensemble that is functionally resilient but collectively impoverished. Whereas I have pointed in other places to the technological means by which such functional resilience is realized, in this chapter I focused on the symbolic and cultural mechanisms whereby the ontological crisis posed by the disorder propagating from the body on the tracks is transmuted and deflected into a routine logic that leaves the commuter suicide recognized but unacknowledged. Thus the dynamic tension at the core of the margin of indeterminacy and collective life—a tension between pattern and disorder, structure and event—is collapsed, leaving only pattern and structure, or repetition without difference. In sum, the commuter suicide marks a limit in the technicity of the commuter train network that nevertheless fails to resonate as a limit.

While it is common in Japan for family, friends, and sometimes even strangers to leave flowers on the site of an accident, flowers are never placed on train platforms to mark a commuter suicide. While it would thus seem that the body on the tracks leaves no material traces, for the workers whose task it is to clean up after a commuter suicide the body's presence is not always so easily expunged. As in Satō's account of the cleanup process, during the act of separating the pulverized body from the technological ensemble, the body as such stages a return as a powerful material force capable of disrupting the logic of recognition. Insofar as that interruption is ephemeral at best, it proves instrumental for thinking with the representational and thematic disorder of the film *Suicide Circle*. I have taken certain liberties in explicating the representational significance of commuter suicide in the film. But is not such labor of explication in the face of disorder precisely what the body on the tracks demands? That such labor is predicated on acknowledgment rather than recognition is clear in the way the film purposely misleads the audience members into thinking they recognize the connections behind the events in the film, only to throw them finally into uncertainty. "You think you know this story, but you do not," the film tells us. To truly understand would be, as the child in the film suggests, to "feel the pain of others as you would your own. . . . bear the concerns of others as you would your own." Such are the conditions of acknowledgment, and, as we see in the next chapter, what is ultimately demanded of a community that suffers a terrible loss of life as a result of a train accident. Ultimately, *Suicide Circle* does not provide any easy answers regarding commuter suicide; instead, it treats the issue of commuter suicide as an exemplary instantiation of a kind of event within an immersive technological environment. As such, the commuter suicide comes to signal the possibility for collective becoming in ways that mobilize technology toward a different degree of technicity.

Finally, it is worth noting that in 2016, Japan's National Police Agency (NPA) reported that, in line with a trend beginning in 2012, the annual suicide rate in 2015 had fallen below 25,000 for the first time since 1997. With that decline, the frequency with which a commuter suicide brings commuter-train traffic to a halt has also decreased. How we are to read this decline remains to be seen.

Notes

1. The website is now defunct. My last access was in June 2010. The address was http://www.h5.dion.ne.jp/~lilith13/jisatu4.txt.
2. Sato, *Jinshin jiko dēta bukku 2002–2009*, 3.
3. Prior to the introduction of the term *jinshin jiko*, it was not uncommon for newspapers to carry reports of individuals committing suicide by jumping in front of the train (*tobikomi jisatsu*). Such accounts also tended to provide details of the jumper's life and references to motives such as mental illness, employment problems, or relationship problems. These details disappear entirely in reports of *jinshin jiko*. For a further discussion of this, see Fisch, "Tokyo's Commuter Train Suicides and the Society of Emergence."
4. I rarely encountered a commuter able to parse the term *jinshin jiko* in terms of its official meaning.
5. Until 2015 the annual suicide rate has stayed above 30,000 (see www .mhlw.go. jp/toukei/saikin/hw/jinkou/tokusyu/suicide04/2.html). Exactly how many of the train-service disruptions between 1989 and 2000 were the result of commuter suicides is difficult to determine from annual statistics published by Japan's Ministry of Land, Infrastructure, and Transport, which does not provide region-specific data. An independent publication provides more detailed figures and precise classifications for the period of 2002–2009, listing an increase from 85 to 169 suicides on train lines within the greater Tokyo metropolitan area (Sato, *Jinshin jiko dēta bukku 2002–2009*).
6. For an in-depth scholarly treatment in English of suicide in Japan from the medial and social perspectives, see Kitanaka, *Depression in Japan*. When, but also if, Japan's recession came to an end is a topic of debate. Some place the end as late as Prime Minister Abe Shinzo's return to government in 2012 and his formulation of "Abenomics." Critics of Abenomics, however, point to its less than spectacular performance and the continuing recession.
7. The comment appears in the reader response section to an article titled "Hinpan ni aru densha no jinshin jiko. Waruii no wa dare?" [The frequency of *jinshin jiko*. who is to blame?], *Nyūsu batake*, November 26, 2007, accessed August 23, 2011, http://news.goo.ne.jp/hatake/20071119/kiji228 .html. I also cite this article in Fisch, "Tokyo's Commuter Train Suicides and the Society of Emergence."

8. For examples of such theories of a uniqueness to Japanese suicide, see Iga, *The Thorn in the Chrysanthemum*; Pinguet, *Voluntary Death in Japan*; Shneidman, *Comprehending Suicide*; Farberow, *Suicide in Different Cultures*; Headley, *Suicide in Asia and the Near East*.

9. See Kitanaka, *Depression in Japan*; West, *Lawn in Everyday Japan*; Takahashi, *Chūkōnen jisatsu*.

10. West, *Law in Everyday Japan*, 215.

11. Douglas, *Purity and Danger*, 36–41.

12. The Deleuzian reference here is intentional, as the argument will return to his text. See Deleuze, *Difference and Repetition*.

13. See Thomas Lamarre's explication of technological optimization in LaMarre, *The Anime Machine*, 137–38.

14. As Muriel Combes's work suggests, Simondon opts for the term *reticularity* in place of *network* to underscore the element of a mutual becoming: Combes, *Gilbert Simondon and the Philosophy of the Transindividual*, 66–69.

15. "Jinshin jiko asa no JR Chūō sen 11 man-nin ni eikyō" [*Jinshin jiko* on the morning Chūō Line 110,000 people affected], *Mainichi Shinbun*, August 19, 2004.

16. Shimabukuro (commuter), interview by the author, August 19, 2004, at Higashi-Koganei Station, Tokyo.

17. Deleuze, *Difference and Repetition*.

18. Osborne, "'Fascinated dispossession.'"

19. Ibid., 282.

20. Iida, *Rethinking Identity in Modern Japan*, 1.

21. See David Wilberry's discussion of agency, body, and culture in the foreword to Kittler, *Discourse Networks 1800/1900*.

22. Takahashi, in conversation with the author, August 20, 2004, at Higashi-Koganei Yakitori (tavern).

23. Man at Higashi-Koganei Yakitori (tavern), in conversation with the author, August 20, 2004.

24. Takahashi, conversation.

25. JR East public-relations representative, interview by the author, January 18, 2005, JR East Central Office Building, Shinjuku, Tokyo.

26. This is according to statistics for 2002–2009. Sato, *Jinshin jiko dēta bukku 2002–2009*, 182.

27. Kitanaka, *Depression in Japan*.

28. Satou Tarou, "Gojyussai [shorō utsu] no norikirikata: yon nin tobikomi jisatsu" [How to get over fifty-year-old middle-age depression: four people commit suicide], *Aera Weekly Magazine*, December 7, 1998, 61.

29. Kurosawa, *Tokyo Sonata*.

30. JR East public-relations representative, interview.

31. Between 2002 and 2009, the number of confirmed suicides on JR East train lines rose from 92 in 2002 to 149 in 2009. Sato, *Jinshin jiko dēta bukku 2002–2009*, 194.

32. Bataille, *The Accursed Share*.

33. In using the term *investment without return* I am borrowing language from Shershow, *The Work and the Gift*.

34. Bataille, *The Accursed Share*, 26.

35. "Ekichōsan 'kamidanomi' jisatsu tsuzuki, kyūyo JR chūō sen Tokyo" [Stationmasters "entreaty to deities" suicides continue, desperation on Tokyo's Chūō Line], *Asahi Shinbun*, December 1, 1995. *Oharai* is often translated as a Shinto purification ritual and carries the sense of appeasement of the spirits.

36. Similarly, it is customary for people in Japan to visit a shrine for *oharai* at times to expunge themselves of impurities that they have collected in everyday life without knowing; such impurities, if left untreated, could be the source of future misfortune.

37. "Chūō sen de jinshin jiko 'oharai' kōka nashi, Tokyo" [Human accident on the Chūō Line "entreaty to deities" useless, Tokyo], *Asahi Shinbun*, December 19, 1995.

38. Deleuze and Guattari, *A Thousand Plateaus*, 256.

39. Satō, *Tetsudō gyōkai no ura banashi*. For the blog version, see http://railman .seesaa.net/article/31413915.html (accessed March 4, 2017).

40. Ibid., 13–14.

41. Ibid., 19.

42. Ibid., 21–22.

43. Satō Mitsuru, interview by the author, September 1, 2012, in Shinjuku, Tokyo.

44. Satō, *Tetsudō gyōkai no ura banashi*.

45. For a discussion of this in the United States, see Turkle, *Simulation and Its Discontents*, and Turkle, *Alone Together*. For the discussion in Japan, see Miyadai, Fujii, and Nakamori, *Shinseiki no riaru*.

46. In English subtitles for the film, the band's name is mistranslated as "Dessert." Its correct spelling as Desert is made clear in the book *Jisatsu sākuru*. Sono, *Jisatsu sākuru* (Kadokawa-Daiei Pictures, 2002). The book, which tells a related but quite different story than the *Suicide Circle* film, is presented in the film *Noriko no shokutaku* (2005).

47. Sono, *Jisatsu sākuru*, 04:46–05:28. My translation.

48. Ibid., 1:11:34–1:13:25. My translation.

49. Ibid., 1:13:25–1:13:54. My translation.

50. Pinguet, *Voluntary Death in Japan*, 27.

SIX

Ninety Seconds

April in Japan is a time of new beginnings. It is when the academic year and the fiscal year begin and when the cherry blossoms bloom. As the flowers emerge from their buds, city parks and tree-lined boulevards become lively scenes, packed with new students, new company employees, and groups of young and old sharing snacks and drinks on plastic tarps and blankets spread under the dense canopies of delicate pink and white cherry-flower petals. By late April, the energy and excitement that accompany these events begin to wind down, and life settles into the daily routine of the new academic and fiscal year. It was during this time, at 9:18 a.m. on the clear spring day of April 25, 2005, that a packed seven-car commuter train derailed near a station just outside Japan's second-largest city, Osaka, and slammed into a nine-story apartment building. The impact left the first railcar compacted into the building's ground-floor parking area—a space half its size—and the second railcar broken in the middle and wrapped in a < (*ku*) shape around the corner of the building. The third and fourth cars came to rest on a diagonal across the tracks. After three days of rescue operations, 107 bodies, including the driver's, had been pulled from the twisted wreckage of shiny aluminum carriages, while hundreds of injured passengers remained hospitalized.[1]

The derailed train was on the Fukuchiyama Line, which is owned and operated by JR West, one of the five JR passenger-train companies formed from the privatization and breakup of Japanese National Railways in 1987. The Fukuchiyama Line passes through older city suburbs northwest of Osaka on its way to more recently established exurbs (bed towns) in the hills. Because the train line also runs close to a number of high schools and universities, many of those killed and wounded were students. Although it would be years before JR West and the Ministry of Transportation released actual data from the accident, it was eventually learned that the train had been speeding at 116 kilometers per hour on a section of track designed for a maximum speed of 70 kilometers per hour. According to the owner of a small spindle-machine shop by the tracks who witnessed the derailment, the train went by so fast that he almost mistook it for the Shinkansen, Japan's super-high-speed intercity "bullet" trains.[2]

Scandal followed the tragedy. The first television-news reporters to arrive on the scene dutifully conveyed JR West's suspicion that the

FIGURE 6.1. April 25, 2005. The Amagasaki Derailment.
Source: Kyodo News.

derailment was caused by a rock on the track, with cameras showing
JR West employees carefully examining the railbed at the entrance to
the curve.³ Within hours, however, the focus of reports in the media
shifted to questions concerning the speed of the train at the time of
the derailment. The driver, it was learned, had overrun the stop at the
station before the accident and had had to back up. First reports mentioned an overrun of eight meters, but witness accounts later put the
number at anything between sixty and a hundred meters. The mistake
had cost the driver a precious ninety seconds, which he had apparently
been racing to recover when the train jumped the track on the curve
just outside Amagasaki Station. Sensing a vulnerable corporation and
a lucrative spectacle, Japan's investigative-media machine got to work.
Over the course of the next few months, the accident unfolded as an
ongoing media item, and journalists and authors dug up evidence of
JR West's negligence, mismanagement, and abuse of its employees. In
order to cut operation costs, it was revealed, JR West had not installed
safety mechanisms developed in the 1960s that would have warned
the driver of excessive speed or even automatically slowed or halted
the train. Forty members of JR West management, it was also reported,
had proceeded with a bowling tournament on the day of the accident
and simply watched rescue operations unfolding on the bowling alley's
television. The bowling tournament was followed by a night of drinking at a restaurant-tavern, where members of the party tried to make
certain the company name did not appear in the guest register in order
to avoid attention—a ploy that obviously failed. On the following day,
JR West held a golf tournament, and the next week, members of management went on a holiday trip to Korea. The media also discovered
that two JR West train drivers en route to work had been on the derailed train and had immediately called in to report the accident, only
to be told to continue to their jobs as usual and not to be late.

But by far the most damaging revelation in the media was that,
one year prior to the accident, Takami Ryūjirō, the twenty-three-year-
old driver of the derailed train, had been subjected to thirteen days of
humiliating physical punishment and psychological abuse for a similar overrun error. Physical and psychological abuse, it turned out, was
standard punishment in JR West for any error, no matter how slight.
Glossed as "day-shift (re)education" (*nikkin kyōiku*), the typical battery
of punishment included menial and demeaning labor (pulling weeds,
cleaning bathrooms, polishing rusty rails), copying out the company's
rule and safety book by hand, and handwriting and revising countless
"statements of confessional regret" (*hansei bun*). Individuals undergoing

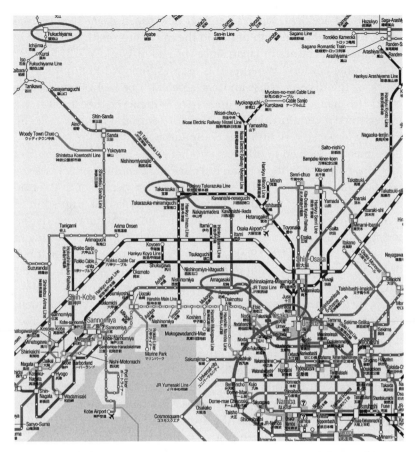

FIGURE 6.2. Map of Kansai train network. Stations on the Fukuchiyama Line are circled.

day-shift (re)education were not permitted to use the bathroom without permission and were typically confined to a single room throughout the day except when performing menial tasks. The story of Takami's punishment prompted revelations from former drivers about similar treatment they had received for even minor infractions, such as being a minute late to a meeting or causing a few seconds' delay in the departure of a train.[4]

Gap

The Amagasaki derailment or Fukuchiyama Line derailment accident, as it came to be known, was certainly not Japan's first nor even its most deadly train accident. Japan's national newspaper, *Asahi Shinbun*, was quick to remind the nation of this in the initial aftermath with a list of all major train crashes in Japan since the end of World War II featured in its online edition.[5] Whether this list was intended to diminish the shock of the disaster ("just another accident") or fuel outrage ("how could this happen again!") is uncertain. Whichever the case, the list underscored a certain troubling familiarity surrounding the accident. In many respects, very little was really surprising about the Amagasaki derailment. It seemed to offer yet another tragic repetition not only of previous train accidents in Japan but also of any number of devastating technological accidents of the past century. In respect to both the conditions of its making and the scandals and conflicts that erupted in its wake, the Amagasaki derailment could, with minimal interpretative effort, be put in the same historical column as the mercury poisoning in Minamata, Japan, in the 1950s; the gas leak in Bhopal, India, in 1984; or the Exxon Valdez oil spill in the Gulf of Alaska in 1989, to offer just a few examples from a depressing list of technological disasters.[6] Similar to these other technological accidents, the Amagasaki derailment presented a story of industrial operation according to the profit motive, a story of corporate negligence fueled by a toxic corporate culture, and a story about the forces of privatization. In its aftermath, it gave rise to the production of conflicting truths between the corporation and the state on the one side and the accident victims on the other side. What was specific to the Amagasaki derailment was the gap— the ninety-second delay that the twenty-three-year-old train driver, Takami Ryūjirō, opened when he overran the stop at Itami Station.

In the aftermath of the accident, the gap was at the center of a controversy between JR West and the "4.25 Network," a grassroots organization of accident victims representing the commuter community. The ways in which JR West and the 4.25 Network understood and treated the gap could not have differed more. JR West maintained that the cause of the accident was human error, specifically the driver's failure to close the ninety-second gap in a safe manner. Investigators from the state's Ministry of Land, Infrastructure, and Transport largely concurred with this view. By contrast, the 4.25 Network held that the cause of the accident was irreducible to a single determinant. They insisted instead that attention be given to the conditions of possibility behind the accident, which they identified as a *daiya* without *yoyū*, or leeway. How, they wanted to know, had such a system come about in which a mere minute-and-a-half delay could end in such a tragic loss of life? Insofar as members of the 4.25 Network and community blamed

JR West for the accident, they were also ready to reflect on their own complicity in engendering a tempo of life and an institutional system so unforgiving that they could not accommodate a ninety-second gap.

In treating the accident as the result of human error, JR West aimed to reduce the gap to an incidental aspect of regular train operation that could be remedied through standard technological fixes and organizational adjustments. In other words, JR West sought to close the gap. By contrast, the community sought to hold the gap open. For the community, the gap was an unmapped and indeterminate terrain that compelled reflection not just on the accident but on the values that had produced the techno-social environment of postwar Japan. These different approaches to the gap informed the manner in which each party sought to process and understand the lessons of the accident.[7] As members of the community were quick to point out, for JR West, redress seemed to mean following a received playbook of accident management requiring nothing more than public apology, a promise of technological fixes, and some cosmetic structural reorganization. Even the severity of the scandals following the accident did not appear to alter JR West's approach. By contrast, for the commuter community and especially for the victims, there were no ready-made models or narratives through which to process the accident. An understanding of the accident could only be generated, they contended, through an exploration of the gap, which led them to explore different modes and registers of expression. Most importantly, it led them to think critically about the problems of trust and complex technologies.

Trust remained a central issue long after the dust of the accident had settled and JR West had managed to resume service on the train line. Time and again, the community rejected JR West's attempts to earn back its trust with claims that it had adequately grasped the lessons of the accident and addressed safety issues on the train line. In this rejection, I argue, the community expressed not only deep mistrust of JR West but also, more importantly, a desire to reconceptualize the nature of trust in relation to the idea that so-called "expert systems," like transportation networks, require individuals to surrender themselves to a certain level of risk.[8] My discussion in this chapter follows the arc of this reconceptualization. The community, I argue, comes to rethink trust through the gap as an integral part of the always unfinished process of collective-making constituted around a margin of indeterminacy. Trust, in this regard, comes to exceed its psychological connotations to take on the quality of a material force of collective coherence. As such, it compels one to ask not only whether a certain person or institution can be trusted to safely operate complex technology, but also whether a technology can be trusted. To trust a technology requires making space for it in a relationship. Trust in a technology demands the willingness to acknowledge machines as partners of collective order that are irreducible to an absolute and determined performance. This is not just about recognizing error or malfunction as inevitable. It is about encouraging the development of machines that have the *yoyū* for human society. The Amagasaki derailment elicits such an orientation toward trust and, in so doing, gestures toward an intervention into a considerable body of literature that has grappled with the relationship between risk, trust, and the inevitability of failure in complex technological ensembles. This gesture became particularly salient less than a decade later in the aftermath of the accident at the Fukushima Daiichi nuclear-power plant in northeast Japan, which catapulted issues of trust and the safety of complex technological systems once again into the spotlight.

The process of thinking the accident with the gap developed over the course of almost a decade between 2005 and 2014; it encompasses a series of official and unofficial responses to Japan's Aircraft and Railway Accidents Investigation Commission's 2007 accident report. The initial moments of this process, however, emerged from within events, memorials, and ceremonies organized by members of the commuter community in the first year after the accident. I use the term *community* here and throughout this chapter to refer to the population living in the immediate area affected by the accident. That the accident occurred within Hyogo Prefecture and not central Osaka is important in this regard, because neighborhoods in Hyogo tend to maintain a community network. JR West's response to the accident and the ensuing scandals galvanized these existing communities, giving rise to the 4.25 Network. My access to and understanding of this community and its dominant sentiments were greatly facilitated by my in-laws, who have lived close to the Fukuchiyama Line for most of their adult lives. Before the Amagasaki derailment, I rode that same train numerous times while staying with them. After the accident, my in-laws eagerly contributed to my research by introducing me to their friends and acquaintances who had been on the train or had lost loved ones in the tragedy. They also introduced me to people from the area who were longtime JR West commuters. Invariably, upon returning from these interviews, I discussed what I learned with my in-laws, sometimes late into the night. My father-in-law's knowledge was particularly instrumental in informing my understanding of the issues. As a retired civil servant who had worked for almost his entire life in the local city hall in various roles of governmental administration, he was especially adept at interpreting and explaining the complexities behind the tension between JR West and the community. In the years following the accident, he continued to send me, often via airmailed packages, a steady stream of newspaper articles he had diligently clipped from local prefectural or community newspapers. In contrast to the scandal-driven spectacle pursued in the mainstream national media, these local articles were typically carefully articulated, thoughtful, and even exploratory analyses of the accident and its meanings. All of this material, as well as the many discussions I had with my in-laws and the people to whom they introduced me, contributed to this chapter. It inflected as well my reading of various published analyses of the accident, from journalistic investigations to government accident reports and local responses.

Trust

In highlighting a problematic of trust, the Amagasaki derailment provides an intervention into thought about technology and risk management. The literature on risk management and technology is extremely broad. While it is not my intention—nor does it fall within the scope of this chapter—to present an exhaustive analysis of this work, I want to draw attention to some of the seminal moments and discussions that have structured thinking regarding risk and technological accidents. Risk has, of course, been a concern for economists and in economic theory for some time.[9] The German sociologist Ulrich Beck, with his notion of the "risk society," first systematically parsed the relationship between risk and technological failure as a social concern.[10] Beck's work, while subject to significant critique, has been the generative seed for a broad and interdisciplinary body of literature often organized under the rubric of risk studies in the United States and Europe. In Japan, risk studies tends to be included under "failure studies" (*shippai gaku*).[11]

In Beck's thesis, risk society is an epochal categorization denoting a qualitative and quantitative shift in the nature of risk incipient with the onset of industrial modernity. With the emergence of second-nature technological environments in conjunction with industrialization, Beck argues, we see an increase in existential uncertainty as a result of the emergence of "manufactured risks" whose impacts are potentially more global and catastrophic than preindustrial and non-technologically manufactured risks. The Chernobyl nuclear accident is exemplary in this regard for Beck. Key to Beck's risk-society thesis is the idea that, with the onset of increasingly severe manufactured risks, the state is less able to manage the certainty of everyday life. Consequently, risk becomes "distributed" and "individualized" across the population as citizens are forced to incorporate practices of risk management into their own daily decision-making processes. Such necessity engenders the emergence, for Beck, of "reflexive modernity," which denotes the rise of a risk-savvy public, able to assess and critically navigate competing expert-knowledge claims regarding risk.[12] Embodying to a great extent Jürgen Habermas's ideal of a public comprising highly rational and scientifically driven individuals, the subject of Beck's "reflexive modernity" thinks about trust in purely instrumental terms, as a kind of currency that institutions accumulate in accordance with their rational, accountable, and competent performance. In other words, trust, in Beck's thinking, becomes a mechanism of rational life management. Elaborating on Beck's thesis, the British sociologist Anthony Giddens argues that in the age of the risk society, the public's trust in expert systems must rest on an empirical assessment of a technology operator's performance and on faith in the systems of rational bureaucratic governance that oversee industry standards and experts. Such trust, for Giddens, is of an impersonal and rational nature, in contrast to the kind of trust that is built in personal one-to-one relationships, which he characterizes as prevalent within nonmodern societies.[13]

Scholars have taken issue with the underlying bureaucratic rationality on which Beck's and Giddens's notions of trust rely. The STS scholar John Downer, for example, argues that risk in complex technological systems cannot be mitigated through the mere enforcement of regulatory compliance.[14] Risk management depends, he suggests, on officially sanctioned informal mechanisms that exploit precisely the kind of personal one-to-one relations that Giddens relegates to nonmodern societies.[15] Offering a different kind of critique, Sheila Jasanoff calls for "democratic risk governance" that takes risk assessment out of the hands of so-called scientifically informed rational actors and puts it under the care of the democratic politics of a population attuned to the nuances of cultural differences.[16] What I identify in this chapter as thinking trust with the gap resonates with Downer's and Jasanoff's arguments in that it rejects the rational conceptualization of trust put forth by Beck and Giddens in favor of a trust founded on close, informal relationships and community governance. At the same time, thinking trust with the gap moves beyond Downer's and Jasanoff's interventions in that it emerges with an understanding of the accident as a result of a system without leeway. I do not mean simply that the system was organized around a *daiya* demanding absolute precision. When the community affected by the accident criticized JR West's creation of a *daiya* without *yoyū*, they were referring to the problem of an absolutely unforgiving system, a system without latitude. JR West's unforgiving system, in this regard, emerges through a disavowal of the presupposition that relationships, whether between human and machine or between human beings, are co-constitutive. It operates instead according to an ideological premise that form determines matter. This premise becomes manifest in JR West's insistence that train drivers make their machines submit to their will, as well as in such practices as "off-train (re)education training" in which human error is treated as the result of a weak will. On yet another level, this premise leads JR West to assume the commuter community will submit to its authority and passively accept its claims of trustworthiness.

This notion of thinking trust with the gap refers not only to processes of reflective thought elicited by the gap but also to thought that takes the gap itself as a condition for thinking trust in relationships. The gap, in this regard, resonates as an ontological force. It informs an understanding of trust as a relationship that is constituted within the space that is opened in the act of forgiving someone for something that is unforgivable. This conceptualization of trust emerges in the aftermath of the Amagasaki derailment within the gestures of bereaved parents and spouses toward JR West. What is more, it opens an avenue for thinking about a kind of *technological* trust, not just trust in human relationships. In so doing, it provides an initial response to another direction of thought in technological-risk studies that claims technological accidents are to a certain extent unavoidable, and thus absolute safety is an impossible ideal. This claim arises from a critique of Jasanoff and Beck for assuming that technological accidents are simply the result of engineering or organizational flaws that can be designed or "organized out of existence" through the imposition of more rational or democratic modes of management.[17] Certain technological accidents, the argument goes, are the result of epistemic limits and are thus entirely outside human agency.[18] In other words, certain kinds of accidents are simply inevitable. One of the seminal concepts in this thesis is Normal Accident Theory (NAT). Developed by the Yale sociologist Charles Perrow, NAT maintains that catastrophic machine failure can result from an unpredictable convergence of irregularities that are otherwise entirely within acceptable parameters of divergence for the normal operation of a complex system.[19] Put differently, failure is an emergent phenomenon and thus beyond intervention, be it through organizational planning or technological design. In an approach that follows close on the heels of Perrow, Downer works from a constructivist understanding of knowledge developed in STS to emphasize the persistence of irreducible ambiguities in engineering knowledge and tests. Machine failure, Downer concludes, does not have to result from normal operational irregularities (NAT) at all, but rather may result from the "unknown unknowns" of engineering knowledge.[20] What Downer calls "unknown unknowns" corresponds in many respects to the Japanese term *sōteigai*. Composed of the Kanji *sō* (concept, think, idea, thought), *tei* (measure, determine, establish), and *gai* (outside), the three-character compound denotes the occurrence of an event or phenomenon that is beyond expectations calculable using existing risk-management models and diagnostic technologies. *Sōteigai* can refer to unforeseeable catastrophic events within a technological ensemble or technological accidents resulting from unforeseeable acts of nature. JR West tried at one point to argue, for example, that the Amagasaki derailment was *sōteigai* because the driver's speed on the curve was an unforeseeable human error. More importantly, less than a decade after the Amagasaki derailment, TEPCO, the operator of Fukushima Daiichi nuclear-power plant, would claim that the reactor meltdown at its plant following the earthquake and tsunami on March 11, 2011, was *sōteigai* because they could not have anticipated a tsunami so massive that it would overwhelm the facility's seawall defenses.

My exposition of the Amagasaki derailment and of thinking trust with the gap is organized into roughly three sections. The first section provides a view of the historical background of the accident and a picture of events and actions leading up to the derailment at 9:18 a.m., with particular attention to the train driver and the question of human error. In my description of the latter, I rely on details and analyses provided by the official 2007 *Railway Accident Investigation Report*, not merely for the rich description it offers but, more importantly, because the report and its conclusions became the focus of the community criticism that I turn to in the second section. That criticism, I find, is articulated especially in the events and speeches that raise the problem of trust around the one-year commemoration of the derailment. Finally, the last section explores how this discussion of trust continued in the decade following the accident in a number of published reports from study groups organized under the initiative of the community and in which JR West participated.

JR West's Urban Network

How does an unforgiving system come to be? Or, to put it differently, what are the conditions of possibility in which a *daiya* without *yoyū* becomes thinkable? To ask this question is to interrogate not just the conditions behind the Amagasaki derailment but also the history of a received culture of techno-politics predicated on the twin notions that the ontological boundaries between the human and the nonhuman are impermeable and that machines are simply instruments of human will. Such a culture of techno-politics was behind JR West's campaign to develop an "urban network" (*āban nettowāku*) that would provide greater speed, convenience, and connectivity for commuters throughout Osaka, Kobe, and Kyoto, the three urban centers of the Kansai region. JR West embarked on its urban network project in 1987, following the company's formation as a result of the privatization and breakup of the Japanese National Railways in 1987. The Amagasaki derailment, however, cannot be explained as the effect of privatization and (neoliberal-esque) deregulation of the transportation system. Privatization was just one of many forces that contributed to the conditions behind the accident, and its significance was a matter of equivocation. While many within the commuter community blamed privatization for encouraging a prioritization of profit over safety, just as many saw the accident as a result of the residual effects of incompetent management from the long era of nationalized railways under JNR. What is more, private rail companies have long dominated the transport industry in the Kansai region, so much so that during the era of JNR the region was called "Kingdom of Private Rail" (*shitetsu no ōkoku*).[21]

JR West's main competitor in the region is Hankyū Railway, which operates train lines around the same three urban centers. The Hankyū brand is deeply embedded in the cultural history of the area, with the Hankyū company operating a major upscale department store. In 1910, following an early-twentieth-century model for closed ecologies of production and consumption, Hankyū Railway laid a train line connecting its Hankyū department store in central Osaka with the resort-spa town of Takarazuka in the northwest hills. In 1913, it went on to establish its world-famous all-female musical-theater troupe in Takarazuka in order to bring in more consumers.[22] The look, feel, and experience within the Hankyū system corresponds with this emphasis on leisure; thus, in many ways, Hankyū embodies the opposite of the look and feel of the JR West system with its emphasis on commuting. Hankyū trains are much heavier, with thicker sides that convey a sense of solidity and safety. Its train cars also carry far fewer advertisements. I have heard riders describe the conditions in Hankyū trains as "less noisy" than the advertisement-inundated JR West train cars. Hankyū platform attendants also tend to bow as the doors of the train close and the train prepares to depart.[23]

Hankyū Railway and JR West's networks exist as separate systems on a shared terrain, albeit with a number of places where commuters can walk a short distance between stations to switch systems. Both train lines depart from hub stations in central Osaka that are located only a few hundred meters from each other. Both train lines then cross the Yodo River on their way to Takarazuka in the northwest hills. The first major station on the Fukuchiyama Line after the Yodo River is Amagasaki. From there, the tracks curve sharply north, passing through Itami before turning to the northwest again just before Kawanishi. At Kawanishi a long and partially covered pedestrian walkway connects the JR West Kawanishi-Ikeda Station with the Hankyū Railway Kawanishi-Noseguchi Station. The JR West and Hankyū Railway lines then run parallel for another short stretch up to Takarazuka, where commuters can again switch between systems if they choose.

Especially intent on winning over commuters from Hankyū, JR West worked to optimize the performance of its network through a series of technological and operational modifications to the Fukuchiyama Line. In 1989, it began implementing its optimization plan with *daiya* revisions that cut travel time between Takarazuka and JR West's Osaka hub. By 1992, JR West was fifteen minutes faster than Hankyū between Takarazuka and Osaka during rush hour, and ten minutes faster during regular hours. But as ridership increased on JR West's train line as a result of real-estate development near its Takarazuka Station, conditions in its rush-hour trains also worsened.[24] JR West thus increased the maximum driving speed on the line from 100 to 120 kilometers per hour, and in 1993 it introduced new, lighter train cars with compressed-air suspension systems to allow for higher speeds on straightaway tracks and on curves. While the trains offered a new seat design with improved comfort, the light construction would later be blamed for providing no protection in the Amagasaki derailment, as the metal was said to have become razor sharp when torn, cutting through the bodies of commuters.[25]

In the fall of 1997, JR West redesigned the tracks near the Amagasaki station, reducing the radius of the six-hundred-meter curve by half in order to allow commuters arriving from the suburban areas around

Takarazuka and a series of new development towns ("bed towns") in the northwest to transfer quickly for connecting train lines to Kobe and the Shinkansen hub at Shin-Osaka Station. Despite these efforts, JR West still struggled to attract riders from the competing Hankyū Line. Commuter allegiance to Hankyū was (and remains) so strong, in fact, that commuters starting their commute on JR West's Fukuchi-yama Line often chose to walk a few hundred meters between the JR West and Hankyū stations at Kawanishi in order to switch over to the Hankyū Line.[26] But beyond a sense of allegiance, the considerable price difference also compelled them to switch from JR West to the Hankyū system at Takarazuka or Kawanishi. As a result of a number of compli-cated pricing schemes carried over from the JNR era, it was actually cheaper for commuters coming from towns further northwest of Ta-karazuka to begin their commute on JR West's line and then change over to Hankyū's train line at Takarazuka.

Having reached an initial limit in possible technological modifi-cations but still intent on winning over Hankyū commuters, JR West initiated a series of aggressive changes in the *daiya* between 2003 and 2004 in order to shorten the running time between Takarazuka and Osaka by seconds. At the same time, it added an extra stop to its Osaka-bound express train. Yet instead of amending the overall travel time to reflect this change, JR West revised its *daiya* to reduce the existing twenty-second dwell time (stopping time) at major stations to a mere fifteen seconds. The resulting *daiya* was cited by some at the time as an exemplary model of efficiency.[27] Such efficiency, it would become clear in retrospect, was a flawed articulation of a formalistic techno-logical optimization realized by means of an unforgiving *daiya* pred-icated on the assumption that commuters would submit to an abso-lutely determined order. For the railroad analyst and writer Kawashima Ryōzō it was a "dense *daiya*" (*chūmitsu daiya*) that aspired to an "ar-tistic" or "divine" (*kamiwaza-teki*) order of complex and closely coor-dinated connections.[28] In retrospect, it would also become clear that the untenability of this "divine" formalism was foreshadowed in the daily announcements, made at stations, that emphasized the detrimen-tal effects of commuter noncompliance on the system rather than em-phasizing safety: "*Presently, passengers have been rushing to board. This is dangerous and, moreover, causes schedule delays. Please cease immedi-ately.*"[29] Commuters were not the only ones pressed to submit to an un-forgiving *daiya*. Drivers were pressured to push their machines to the threshold, accelerating to speeds in excess of 120 kilometers per hour on straightaways and performing last-second braking at curves and stations.[30] JR West treated a driver's failure to master this "divine" command as a failure to assert mind over matter, an indication of a weak will, and subjected such a "failing" driver to the humiliating physical and mental abuse of off-train (re)education training to reshape the driver's resolve.

Human Error

At the time of the Amagasaki derailment, Japan's Aircraft and Railway Accidents Investigation Commission, under the Ministry of Land, Infrastructure, and Transport, was responsible for investigations of train accidents. In line with protocol, just hours after the accident, the head of the commission assigned a committee to investigate the causes of the derailment. As expected, the committee took its job seriously. Over the course of the next year and a half, it analyzed the voice and instrument recorder, the distribution of debris at the accident site, the train car and track conditions, braking conditions, and the shifting of the center of gravity of the train under different speed and load conditions. It also performed derailment simulations and conducted interviews and surveys among passengers from the train, JR West drivers, and conductors. A draft of its final investigation report was published in December 2006 and circulated among thirteen individuals in preparation for a hearing on the report in February 2007. Among those asked to review the draft were a number of top JR West management employees, a representative from the JR West Labor Union, and a number of academics in human sciences as well as engineering. The final 275-page report was made public in June 2007.[31]

Accident reports are a particular genre of knowledge production. They are supposed to impart the conclusions of "disinterested scientific investigation" by "disinterested investigators."[32] The rhetorical currency of the accident report derives from its capacity to assume a simultaneously reflective, remedial, and impartial analytic stance. The accident report does not pass judgment: it is supposed to deconstruct the accident in order to determine whether it was the result of foreseeable organizational or technological failure or human error. An accident report is thus not part of a criminal investigation. Whether or not a criminal investigation should be undertaken is a separate decision that is pursued subsequent to the findings of the accident report. Typically imbued with the language of technological objectivity, the accident report is also not supposed to engage in politics or ethics. The

Amagasaki accident commission's report foregrounds its claim to impartiality and objectivity, devoting the entire second page to the declaration that "This report regarding the accident is based on the findings of the commission established by the Aircraft and Railway Accidents Investigation Commission to clarify the causes behind the train accident in order to contribute to the prevention of accidents. It was not undertaken to determine responsibility for the accident."[33]

The commission's conclusions would eventually be subject to scrutiny by members of the commuter community, the 4.25 Network, and an independent investigative panel almost a decade later. In particular, criticism from these groups took aim at the report's emphasis on human error within its three-tiered breakdown of "Probable Causes" behind the accident in the conclusion, which read:

Probable Causes

It is considered highly probable [*to suitei sareru*] that the train driver's delay in applying the brake resulted in the entry of the train into a 304 meter radius rightward curved track at a speed of approximately 116 km/h, which was far above the maximum specified speed of 70 km/h, and the running of the train along the curved track at the high speed caused the first car of the train to fall left and derail, which caused the second to fifth cars to derail.

It is conceivable that [*to kangae rareru*] the train driver's delay in applying the brake is attributable to the diversion of his attention from driving the train to (1) listening to the dialogue between the conductor and the train dispatcher by radio communication which was caused by his belief that he had been hung up on by the conductor while he had been talking to the conductor on the intercom to ask him to make a false report and (2) making up an excuse to avoid being put on "off the train retraining" course.

There is a possibility [*kanōsei ga kangae rareru*] that the West Japan Railway Company's train driver management system in which drivers who caused an incident or mistake are put on an "off the train" retraining course that can be considered as a penalty or are subjected to disciplinary action and drivers who did not report an incident or a mistake they had caused or made a false report about such an incident or mistake are put on an even harder "off the train" (re)education training or subjected to an even harder disciplinary action may have (1) caused the driver to make the call to the conductor on the intercom to ask him to make a false report and (2) caused the diversion of the driver's attention from driving the train.[34]

There are no egregious untruths in this summary. The problem lies in its organization of the circumstances, specifically the way it emphasizes human error in the first line with the identification of the driver's failure to brake in time as the leading cause behind the accident. Although the fact that the train was speeding is noted at the end of the first paragraph, it is never pursued. Critics of the report were quick to point out as well that the emphasis on human error here appears predetermined by the mode of inquiry conveyed in the investigation report, which devotes significant focus to illuminating the driver's mental state. Indeed, the report goes into considerable detail regarding the driver's life, much of which is of questionable value to the accident investigation. After specifying such things as his height and weight, it goes on to list his hobbies, his financial state, and his familial and romantic relationships. What is more, it provides a minute-by-minute breakdown of his activity in the days leading up to the accident: what he ate, drank, watched on television, with whom he talked, and so on. Reading through this material, one gets the sense that the commission ardently wanted to link the driver's error to his underlying psychological condition. One can sense the frustration of its authors every time it declares that there was absolutely nothing unusual about Takami. Takami had no serious financial concerns, was not given to depression, and was not a big drinker. He was in a happy long-term relationship and was described by friends and family as having a generally sunny disposition. There was also nothing unusual about his work history. One year before the Amagasaki accident, in May 2004, he had even received his First Class driving license.

In focusing on the driver's mental state at the time of the accident without exploring the psychological effects of his work environment, the report operates on the assumption that the driver's failure to brake in time was an effect of his weak character. This is made especially clear in the second paragraph with the postulation that the driver failed to brake in time because he was distracted by listening to the communications radio, which is implicitly attributed to his insecurity and lack of trustworthiness. Only in the third and final paragraph is the impact of Takami's experience in off-train (re)education training considered. In short, it is not that the summary fails to acknowledge the effects of JR West's punitive disciplinary practices; rather, the problem is that the report relegates that possibility to a third-level determinant while simultaneously hedging in respect to its significance through language that conveys increasing degrees of equivocation: "it is conceivable that" (*to kangae rareru*) and "there is a possibility that" (*kanōsei ga kangae rareru*). In short, the summary can be read as giving tacit approval to JR West's understanding of the accident as the result of human error: Takami's failure to close the gap of ninety seconds.

An Unforgiving System

The summary of "Probable Causes" is particularly bewildering for the way it contradicts much of what precedes it in the report. The report conveys not a tragedy resulting from human error but rather a disaster that arose out of a technological and organizational order without *yoyū*. What I mean by this is not just a system operating under the assumption that the physical world should conform to the dictates of the techno-political economic order, but also a system that is absolutely unforgiving, in which gaps are treated instrumentally as spaces to be closed rather than as intervals of interaction and entanglement.

Takami was without doubt not the most skilled train driver. He had a propensity to become slightly confused, even distracted. Yet his less-than-perfect driving skills are not what caused the accident. It was an unforgiving system in which imprecise performance was treated as a failure of will that transformed Takami's shortcomings as a driver into deadly flaws. On the morning of the accident, Takami had managed with a good degree of competence the highly stressful labor of finessing the interval to accommodate the commuter rush. At the height of the morning rush, he had even succeeded in recovering most of the lost time from a forty-five- to fifty-second departure delay at the first five stations on his route. But just before 9 a.m., as the intensity of the morning rush was starting to diminish, things began to go badly. Arriving at Takarazuka Station (forty-four seconds behind schedule) with an out-of-service train on track 1, he approached the station too fast, traveling at sixty-five kilometers per hour instead of the required forty kilometers per hour. At that point, an Automatic Train Stop (ATS) alarm sounded in the cabin. An ATS is a safety device on the tracks that checks the speed of the train as it passes. In the aftermath of the accident, it became the focus of debate around train safety, as a result of which the public became versed in its specifics through countless exposés in newspapers and magazines. The simplest version of the system activates a bell in the driver's cabin if the train is over the designated speed limit. Developed as a fail-safe mechanism against sleepy, distracted, or incapacitated train drivers, it requires that the driver acknowledge the alarm within a number of seconds by pressing a cancel button. If the driver fails to do so, the brakes automatically activate, bringing the train to a complete stop. More advanced versions of the ATS can automatically slow the train down to the required speed instead. The type of ATS device on the track at the entrance to Takarazuka Station was the simpler model, and Takami did not respond. As a result, the brakes activated, halting the train. Takami manually released the brakes and shut off the ATS buzzer before continuing, only to have the ATS activate the brakes again, which it is designed to do when the reactivation process diverges from a normal pattern.

Why had Takami entered the station twenty kilometers per hour over the speed limit? The commission's report conjectures that he may have temporarily nodded off to sleep but also that he may have mistaken the track he was entering as track 2 instead of track 1. Track 1 curves as it comes into the station and has a maximum speed of 40 kilometers per hour. By contrast, track 2 enters straight, allowing a maximum speed of 65 kilometers per hour. It is thus imaginable that Takami thought that he was on track 2 and thus only a kilometer or two above the speed limit.

Whatever the reason, the report underscores that the incident left Takami clearly unsettled. After bringing the train to a stop in the station, he was supposed to exit the driver's cabin immediately and walk to the rear of the train in order to switch places with the conductor for the return trip to Amagasaki. However, Takami remained in the driver's cabin for an extra three minutes. When he did finally exit and was walking on the platform to the other side of the train, he walked past the conductor, who stopped to inquire what had happened when entering the station. The conductor was forty-two years old and thus nineteen years Takami's senior. For JR West, Takami's impropriety in not responding to his senior coworker was evidence of his lack of professionalism and his unstable mental state. In a detailed and highly insightful analysis of the Amagasaki-derailment investigation, Satō Mitsuru, a former JR East train mechanic and author of several books on commuter-train operations, offers a different explanation for Takami's behavior that illuminates the social organization of a system without *yoyū*.

On the one hand, Satō writes, since Takami and the elder conductor were members of different workers' unions and did not work together on a regular basis, it is possible that Takami simply felt that he did not want or need to respond to the conductor. On the other hand, it is also not clear that Takami meant to ignore him. He may have been lost in imagining the repercussions for his mistake. For not only did drivers have to report the use of the emergency brake and the failure to respond to the ATS, but also JR West typically subjected drivers to severe off-train (re)education training as punishment. Takami had undergone such punishment around one year prior, on June 8, 2004, when he overran the stop at a station by one hundred meters and caused an eight-minute delay. Takami then failed to report the incident immediately. Making matters worse, he was also caught lying when he eventually did report, claiming that he had been braking but just not hard enough when it was clear from the operation recordings that he had not taken any action. The superior to whom Takami reported noted in his report that Takami had a "smirk on his face," which was read as an indication of his arrogance and insufficient regret. The internal investigation into the incident concluded that Takami had most likely been nodding off to sleep in the driver's cab. As punishment, he was assigned to thirteen days of (re)education training and given a warning that any further mistakes of the kind would result in his demotion from driver. Along with the usual regimen of punishment, Takami was forced to write repeatedly during his thirteen days of (re)education training, "If I do it again I will quit" (*kondo yattara yameru*). Leaked to the media, images of the countless white notebook pages filled with his hurried, almost illegible, characters were displayed on the evening news as evidence of the abuse the young driver had suffered. JR West would eventually admit that its off-train (re)education practice was informed by its faith in the idea that with the proper attitude, a driver would triumph over all obstacles. Known in Japanese as *konjōron*, this concept was often invoked to motivate Japanese soldiers in World War II even as the Japanese empire was collapsing.[35] According to JR West, a "professional" driver was one who could mobilize his will in overcoming all material constraints, such as time pressure, work stress, or even equipment flaws. Failure to do so was seen as a sign of a weak spirit and a moral shortcoming.

Takami never quite recovered from the mistake at Takarazuka Station. After departing fifteen seconds behind schedule, he was twenty-five seconds late leaving the next stop, Nakayamadera Station, and thirty-five seconds late leaving Kawanishi-Ikeda. Such delays were of course nothing unusual for Takami. He had managed more-severe circumstances earlier the same morning. Nevertheless, it seems he felt compelled to recover the lost time before arriving at Itami Station. Passing through Kita-Itami Station (a no-stop station for the semi-express), the train was recorded as traveling at 122 kilometers per hour. Less than five hundred meters before the Itami Station platform, it was still traveling at 112 kilometers per hour. It then entered the station far over the speed limit, at eighty-three kilometers per hour. It is hard to imagine what Takami was doing and thinking at that point. After repeated automatic warnings, he finally started braking, bringing the train to a halt seventy-two meters beyond the designated stopping point. Takami then hurriedly reversed the train but missed the mark again by three meters. As a result, the train was one minute and twenty seconds late departing Itami Station, placing doubt on whether it would arrive in time for commuters to transfer at Amagasaki Station for the special rapid express to Kobe. Upon leaving Itami Station, Takami accelerated immediately to 124 kilometers per hour. At the same time, knowing that he and the conductor would have to report the overrun and that he would face another round of off-train (re)education training and most likely demotion, he immediately called the conductor at the back of the train through the intercom to ask if he would help him out by falsifying the report just a bit. The conductor responded, "You certainly overran it, didn't you?" before suddenly hanging up the intercom phone. Considering the fact that Takami had seemingly ignored the conductor on the platform at Takarazuka, one can imagine that Takami became even more anxious, thinking that he had angered the man with his request to falsify the overrun. Since he had been reprimanded severely once already for lying on a report, there was no doubt that another lie would be punished even more harshly. What Takami could not have known, however, was that the conductor had not hung up on him in anger. While the conductor was talking over the intercom to Takami, a passenger knocked on the conductor's window, upset that there had been no apology for the overrun at Itami Station. Feeling compelled to respond, the conductor hung up and immediately apologized in an announcement. He then used the wireless radio to report the overrun to central command. At the front of the car, Takami, it is believed, was listening closely as the conductor lied on Takami's behalf, reporting an overrun of only eight meters—more or less what Takami had asked for. But it was already too late. Moments later the train entered the right curve at 116 kilometers per hour. Takami tried to brake, it seems, but the train was going too fast. At exactly 9:18:54 a.m., the first train car jumped the rail, taking cars two, three, four, and five behind it.

The accident report states that Takami was found with his right glove off and that a special red pencil that drivers use to take notes while operating the train was found beside him. The report conjectures that at the time of the derailment, Takami had removed his glove in order to take notes on what the conductor reported so that he could later adjust his own report. His right hand should have been on the brake.[36]

I have gone into detail here concerning the accident in order to il-

lustrate the character of an unforgiving system. The thread of causal relations leading up to the derailment reflects in many ways the complex
convergence of unfortunate circumstances. The one common feature
among those unfortunate circumstances is their origin in a collective
lacking the empathy and leeway for error. As much as the corporate
culture of JR West and its practice of off-train (re)education stands out
in this regard, one cannot but see the passenger who demanded an immediate apology from the conductor as part of the unforgiving system.
What was to be gained from the apology other than the satisfaction of
his own ego? Was there no other way to imagine the overrun at the station except as an unforgivable error?

Unimaginable

Mika was riding in the seventh (last) carriage of the train on the morning of the accident. "The train always went very fast," she explained,
"but after leaving Itami Station that day it was going so fast that I
couldn't even read the names of the stations we passed."[37] Speaking in
rapid Kansai dialect, Mika relayed her accident experience to my wife
and me two months after the derailment amid the general clamor of
fast-food service at a Wendy's in the basement of the Hankyū Kawanishi Station.

Mika was a slender woman, thirty-one years old with shoulder-
length, fashionably styled hair. At the time of the accident, she was living with her parents and brother in Kawanishi and commuting every
day to work as a paralegal for a law office in Osaka. She considered herself lucky that her job allowed her to avoid the worst part of the morning rush hour. By nine in the morning, the usual crowd of salarymen
and salarywomen typically thinned out, replaced by a livelier group
of high-school and university students along with the odd retiree
trying to get an early start on shopping or events in the city center.
Mika always boarded the last car of the seven-car semi-express train at
9:11 a.m. from Kawanishi-Ikeda station. Although the last train car was
slightly inconvenient in terms of where it placed her on the platform
for the transfer at Amagasaki, Mika preferred it as it tended to be less
congested with students, who always crowded into the front cars in order to be closer to the exit at their stops. In some ways, Mika confessed,
she looked forward to the commute, as it gave her time to be alone before work and indulge in her passion for Edo-period novels. When she
could, though, she tried to coordinate her commute with her brother,

who at the time attended a university in the city. On those days, she and her brother rode together in the first train car, since her brother liked to smoke while waiting for the train, and smoking was permitted only at the front of the platform. The day of the accident, they had planned to commute together, but her brother overslept, leaving her to ride alone in the seventh car. Her brother's penchant for oversleeping had probably saved her life, she admitted, although she would never have thought that was possible. An accident, let alone an accident with so many fatalities, was just not something she imagined. Thus, when the train overran its stop by more than seventy meters at Itami Station (the station before Amagasaki), for Mika it was not a portent of the coming crash but rather an anecdote she thought she might relay later to her friends at work during their lunch break. For those waiting on the platform at Itami as well, the overrun was a strange but ultimately somewhat amusing event. As one woman who lost her mother and grandmother in the accident would struggle to convey through tears at the official memorial ceremony one year later,

The train entered the station so fast it didn't seem at first that it was going to stop. But when it did, it had almost completely passed the station and had to back up, which was strange because it's something one never sees. "This driver must be asleep. Maybe I should go wake him up," my grandmother joked, and the three of us laughed. Moments later the accident happened, and the three of us struggled to hold on to each other as we were thrown. Then I lost consciousness. That's the last time I saw my mother and grandmother alive.[38]

After departing Itami Station, the speed of the train made Mika nauseous. She began to feel that something was deeply wrong. Yet even as the train began to shake violently, the thought of an accident never crossed her mind. So instead of bracing herself for impact, she tried to relax and turned back to the pages of her novel. Moments later Mika felt the train suddenly brake, and she was thrown to the floor with other passengers, injuring her arm. Mika was not alone in being unable to imagine the accident. Of the fifteen victims of the accident who were interviewed by the Aircraft and Railway Accidents Investigation Commission, all recalled sensing something disturbingly off about the commute but only one, a man in his thirties, remembered wondering just before the derailment whether the train could make it through the curve at such a speed.[39]

After Mika and others in her train car managed to stand again, they waited in silence, trapped inside the train, for an announcement from the conductor or for someone from JR West to come help. Neither arrived. When help finally arrived, it was not from JR West, but from workers from a nearby steel-spindle factory. The workers took charge of the scene, improvising with whatever they could find to help the wounded. They ripped the long blue cushions from the train benches to use as stretchers; brought endless bags of ice, towels, and water; comforted the shocked survivors; and commandeered whatever vehicles they could find to begin shuttling the wounded to the hospital.

Equivocations of Trust

At the time I spoke with Mika, service on the section of track where the derailment had occurred had been restored. Service had begun two weeks earlier at 5:00 a.m. on June 19, 2005, in a ceremony that received intensive media coverage. Over one hundred reporters from the press participated in the event, riding on the first train and far outnumbering a small group of twenty or so regular passengers that included family of accident victims. After having dodged reporters during the months since the accident, JR West welcomed the media's scrutiny that morning. There was much at stake for the company, as the derailment and ensuing scandals had completely undermined the commuters' trust in the company's operations. Consequently, despite the relative swiftness with which JR West managed to repair its train line, many commuters were still not ready to ride its trains. For them, JR West needed to do more than simply declare the Fukuchiyama Line open again for business. It needed to recreate trust, which is what it tried to do by turning the resumption of service into a performance and a media event. As the *Asahi* newspaper reported later, at 4:40 a.m. the same morning, JR West's chairman, along with fifty JR West employees, visited the accident site to lay flowers.[40] Later, as the first train departed Takarazuka Station at 5:00 a.m., JR West's president, Kakiuchi Takeshi, stood in a beige dress uniform beside the driver in the driver's cab. When the train neared the infamous curve, it slowed to a mere thirty kilometers per hour while an announcement was made in the train that they were approaching the accident site. As the train passed the now-abandoned mansion, families of accident victims on the train pressed their palms together in front of their chests and lowered their heads in silent prayer. Later, at Amagasaki Station, the article goes on to report, JR West's President Kakiuchi declared as he exited the train that he confirmed the safety of the train line and added that as the train had passed the accident site, he had felt himself tremble. "From the depths of my soul, once again, I apologize," he said. In a matter of months, Kakiuchi was forced to announce his resignation over his negligence in the handling of the accident. He was succeeded by Yamazaki Masao, who would also be charged with negligence but eventually acquitted in 2012.

Mika, like many former JR West commuters, was not prepared to return to the Fukuchiyama Line. She did not trust JR West, she confessed. But her lack of trust did not stem from doubts about the safety of the train. It was JR West's attitude and its relationship with the community that she found troubling. Mika explained that during the two months of discontinued service on the Fukuchiyama Line, JR West had made arrangements for its commuters to use their monthly passes on Hankyū's trains. Just before service resumed, she received a letter from JR West regarding the termination of the arrangement with Hankyū. She described the letter as bureaucratic and perfunctory, stating only that as of the end of June, the JR West commuter pass could no longer be used for the Hankyū Line. "There was no apology for the accident," Mika exclaimed in a tone that conveyed exasperation and disbelief. "So incredibly thick-skinned. Not a single mark of regret regarding the dead, and no expression of concern for the injured or even recognition of the inconvenience it had created!"[41]

Eventually, JR West seemed to understand that it was going to require more than a media event to win back commuters like Mika, and so throughout the next year the company pursued a campaign to earn back the community's trust. At the center of the campaign was a poster series with the theme "To Further Ensure Trust and Peace of Mind" (*sarani anshin shinrai shite itadaku tame ni*), through which the company sought to convey to commuters the various measures it was taking to improve safety and performance. Appearing in trains and train stations on the Fukuchiyama Line, each poster in the series was designated a "volume" so as to lend a sense of gravity and breadth to its content. Over the course of the next decade, JR West would produce forty-five individual "volumes" for the series, all following more or less the same formula.[42] Among these, the first two are the most important for the way they attempt to produce certain truth claims in response to a specific critique leveled at JR West regarding its operations and handling of the accident. In volume 1, JR West addressed the argument that the failures leading to the accident were the result of JR West's corporate culture by announcing the establishment of a new "corporate doctrine" (*kigyō rinen*) and "safety charter" (*anzen kenshō*). In volume 2, it confronted the claim that the accident might have been prevented had safety mechanisms been upgraded and installed before the curve by describing the deployment of new safety devices on the track, with pictures of the devices, technical explanations, and a map of planned installation points for the future.

Overall, the posters tack between two styles of address, presenting on the one hand a confident and active corporate voice that translates easily into catchy slogans and on the other hand a culturally recognizable mode of confessional writing known as *hansei bun*. Volume 1 commences with the latter, stating that the Fukuchiyama Line accident has prompted JR West to reevaluate its corporate character and ethics in order to cultivate a working environment that prioritizes safety. The poster then proceeds to elaborate on this in four separate points, each of which alludes to the idea that the development of the corporate doctrine and safety charter resulted from a collective and organic undertaking that involved the unsolicited active participation of all employees. Images to the right of the text appear carefully selected to embody this collective resolve, with each capturing some recognizable gesture of safety, such as a driver pointing to a signal or maintenance crews examining a piece of technology. The third image carries particular rhetorical force in that it shows workers either examining or installing an ATS train-speed regulation device, which, as mentioned above, JR West was heavily criticized for failing to place at the entrance to the curve before the Amagasaki Station. Below this introductory text, volume 1 presents the corporate doctrine and safety charter in two separate text boxes. Referring to the Fukuchiyama accident as its initial source of inspiration, both the doctrine and the charter adopt a positive, active, and confident tone, emphasizing strong resolve to transform the company's practices. The corporate safety charter addresses JR West's reformed corporate culture by emphasizing the company's devotion to prioritizing the customer/commuter, but it also takes care to emphasize its continuing commitment to increasing company profit, which seems calculated to appease shareholders. It goes on to reaffirm the company's devotion to rigorous safety protocol, careful operation, and observation.

In contrast to the semiconfessional style of volume 1, volume 2 adopts a pedantic and pedagogical tone. It starts by parsing the specific functions of different versions of the automatic train-stopping device, which was already familiar territory for anyone who had followed the story of the derailment in the newspaper. This is followed by a brief report on the current status of work to install devices and a schedule for more planned installations in the coming year. Images of installed ATS devices (clearly demarcated with a red circle) and a diagram of planned installation points work to lend evidence to this report while giving the poster the gravity of a technical exposition. Subsequent volumes in the poster series follow the style of volume 2.

What stands out in JR West's posters is the use of the term *shinrai* for trust. Trust is a complex term in any language and culture. In Japanese, its complexity is compounded by the existence of a second term, *shinyō*. Both terms use the same first character, *shin*, which means

FIGURE 6.3. Volume 1, "Trust and Peace of Mind" Poster series. Courtesy of JR West.

FIGURE 6.4. Volume 2, "Trust and Peace of Mind" Poster series. Courtesy of JR West.

"truth" or "faith." The difference between the terms derives from the second character. Whereas the second character in *shinrai* means "to rely or depend on" (*tayoru*) or "to request" (*tanomu*), the second character in *shinyō* carries the meaning "to utilize" (*yō* or *mochiiru*). The variance in meaning afforded by these different second characters is nuanced but informs the appropriate context of use. Trust as *shinrai* (the term JR West used) underscores relational processes over outcomes. It refers to a sense of trust that emanates from the vital constitution of a specific relationship. To put it differently, *shinrai* needs a face. It is about an interdependence that presupposes a willingness, openness, and readiness to exert the effort needed to maintain a relationship. If I say, for example, that I trust someone (*shinrai shiteru*), my focus is not on achieving a specific desired result from my relationship with them but rather on the feeling that they are open to the process of relationship building. "Peace of mind" is an effect of the sense of trustworthiness that can develop within that relationship.[43] By contrast, the term *shinyō* tends to refer to the credibility and reliability of an institution or thing. Conveying an instrumental connotation, it puts emphasis on positive outcomes. Insofar as *shinyō* can be used to describe the trustworthiness of a person, it stresses function over relationship and thus unlike *shinrai* denotes a quality of trust that does not necessarily need a face. The impersonal and institutional character of *shinyō* makes it similar to the kind of rational trust that Beck and Giddens argue characterizes relations of trust in the "risk society."

JR West appears to have confused the terms *shinrai* and *shinyō* in its posters. In declaring the desire to "earn" commuters' trust (*shinrai*), it was ostensibly expressing its willingness to build a relationship of interdependence with the community. But the posters convey an impersonal and institutional form of trust (*shinyō*) contingent on demonstrating technical expertise, fixing technical problems on the train line, and remedying organizational flaws. In other words, JR West's gesture at trust building was unidirectional. It established new safety measures as an implementation of policy from above in an attempt to prove itself as an institution worthy of trust. At the same time, the posters take on a didactic tone that suggest JR West assumed that, as an institutional authority, its safety measures and promises would be accepted without question. Despite stating their resolve to "earn" trust, JR West's posters try to circumvent the process of building trust, instead pronouncing that JR West should be trusted. In so doing, they convey the company's presumption that reality will conform to policies implemented through a strong will, thus reproducing the problematic disposition that informed JR West's practice of off-train (re) education. Finally, insofar as the posters demonstrate JR West's effort to achieve safe operations, they stop considerably short of guaranteeing safety. The implicit message is that risk can be diminished but not eliminated. Safety is thus treated as a matter of risk assessment, which carries the assumption that most, but not all, accidents can be anticipated and prevented.

In sum, JR West pursued reestablishing trust as a branding campaign in which safety was treated as a commodity. The company objectified trust, rendering it of merely instrumental value. If the posters can be said to have any positive value, it lies in their pedagogical effect. In this regard, though, they merely continued an educational process that Japan's media had started with analyses of the various causes behind the accident. Prior to the accident, most commuters probably could not have explained the differences between different traffic-control mechanisms. After months of countless media exposés in the wake of the accident, however, the commuter community had become railroad savvy, able to parse the differences between an ATS, ATS-P, and ATS-S.

Disconnect

Tension between JR West and the commuter community became increasingly severe the year after the Amagasaki derailment. This tension derived from the fundamentally different approaches each took to the question of the gap. JR West wanted to move past the accident. Having identified human error as the cause of the accident, it wanted to close the gap and move on. The community, however, was not ready to do so. For the community, the cause of the accident remained complex and undetermined, rooted in the cultural, technological, and political circumstances that had allowed a mere ninety-second delay to end in the loss of 106 commuter lives. For the community, the gap was a problem space, not a problem to be solved. It elicited exploration and reflection that community members were prepared to follow toward social change. More than answers from JR West, the community, often led by the 4.25 Network, sought interaction with the company. JR West, however, remained entrenched in its position of authority and unwilling to engage. The result was an atmosphere of disconnect between JR West and the community that became particularly noticeable around the one-year accident commemoration events.

The one-year commemoration commenced at 9:18 in the morning with a fairly large group of people gathered at the accident site to observe a minute of silence. Most of the participants appeared to be regular commuters, male and female, on their way to work. The area around where the train had collided with the apartment building had been sectioned off over the year with makeshift white plastic stretched over a pipe scaffold. A simple structure with an aluminum roof and sliding shutter sides had also been constructed to provide a memorial space, which housed a place to leave flowers (*kenkadai*) and an incense burner (*shōkōdai*). On any given day of the week, it was not unusual to find a line of visitors patiently waiting to pay their respects at the site. On the morning of the one-year commemoration, the lines were unusually long, stretching all the way around the accident site. Despite the number of people, the overall atmosphere remained subdued, and people waited quietly in line and around the tracks. With the announcement that it was 9:18 a.m., they bowed their heads for a minute of silence that was punctuated by the deep and steady chime of a bell from a nearby Shinto shrine. A large number of reporters were also present and the incessant metallic click of camera shutters filled the minute. A number of media crews had even come prepared with industrial boom lifts (or cherry-picker machines) whose long hydraulic arms extended high above the tracks and crowds, swaying back and forth in the north wind that made the otherwise fair-weather day unseasonably cold.

The day before the one-year commemoration, I had spoken with members of JR Sōren, one of the largest railroad workers' unions, whom I had encountered distributing flyers outside Amagasaki Station announcing plans for a major strike the next day. Enumerating the union's grievances against JR West, the flyer mentioned the company's failure to implement a more relaxed *daiya*, its failure to apologize to employees subjected to off-train (re)education, and the continued lack of safety mechanisms on tracks and trains. The youngish-looking and slender male union representative who handed me the flyer seemed visibly excited, if not agitated. He told me that the strike would be a significant event with full media coverage.

The next day, there was no sign of JR Sōren or the kind of confrontational atmosphere intimated in the flyer. But there was also no sign of cooperation or dialogue between JR West and the community. The deep tensions between the two groups over their conflicting approaches to the problem of the gap and trust became especially manifest in the expressions of commemoration and mourning that each had planned for the day. Whereas JR West maintained its paternalistic and aloof stance by keeping to formulaic expressions of grief and commemoration, the community groped for expression. While appearing slightly disorganized, community members were intent on exploring different modes of reflection and communication. This difference in approach was articulated in the choreographic disconnect between the commemorative activities that JR West had organized and those that the community had organized for the main stations between Takarazuka and Amagasaki.

In each of the stations, JR West set up registries for people to sign that would later be presented to victims and bereaved families at

JR West's official memorial service. Marked with a large standing sign that read "Fukuchiyama Derailment Accident Memorial Service," the registry setup consisted of a single table covered in white cloth and two folding chairs, provided for people to use while signing. Two JR West employees stood behind the tables. Dressed in black, with their heads slightly bowed and their hands folded in front, they looked properly solemn. While they remained ready to assist people with signing the registry if needed, they did not interact with them. For most of the day, a short line of people waited patiently for the chance to sign the registry.

In one sense, there was nothing unusual and certainly nothing confrontational about JR West's offering a registry for people to sign. In another sense, that was precisely the problem. JR West was following conventions of commemoration and mourning intended to help produce closure. By contrast, the community did not want closure. Nothing said this more clearly than the slogan that it had chosen for the day: "4.25 A Day We Will Never Forget." In the events and activities that members of the community had planned for the day, they resisted received modes of mourning and appeared ready instead to explore the indeterminacy of the gap through open-ended collective dialogues

FIGURE 6.5. JR West employees at registry.

with no set objective for what they aimed to achieve. Their events were entreaties for collective expression, like the "Coming Together Concert," where people were invited to sing along following music sheets distributed by volunteers. Community volunteers also set up message boards in the five main train stations between Takarazuka and Amagasaki. Designated "Blue Sky Message Boards," the large hand-painted boards stood just under six feet high and three feet wide. In some instances, JR West's registry was a mere few inches away from the message boards, yet they seemed part of radically different worlds. Each set of message boards featured a motif of white doves in flight across a tranquil blue background with the phrases "Time," "So Many," "Love," "Regret," "Encounter," "Friendship," and "Dialogue" written in black, in both Japanese and English.

Distributed across the boards as points in space, these words served as loosely defined message prompts whose connections to each other were explicitly left undetermined, though one could certainly imagine them. The objective, it seemed, was to provide a minimal narrative

FIGURE 6.6. Signing the Blue Sky Message Boards.

scaffold as a fertile lattice for the emergence of an organic expression of grief or even an explanation for why—not how—an ordinary commute had ended so tragically. It was for passersby to produce that connection by filling in the stretches of open sky-blue space between words, and volunteers armed with black pens stood by ready to encourage them to do so. When I first passed the boards at Itami Station on my way to the accident-site ceremony in the morning, there were only a few messages, and the atmosphere around the boards was calm. By the time I returned later that morning, the atmosphere around the boards was vibrant with individuals stopping to read hastily scribbled messages that covered almost all of the blue-sky space.

Memorial Ceremony: Thinking with the Gap

JR West's failure to earn back the community's trust was palpable in the events that the community organized throughout the day to mark the one-year anniversary of the Amagasaki accident. But not until the evening public-memorial ceremony was the matter explicitly foregrounded. Ostensibly, the ceremony was intended as a chance for the community to express solidarity. Organized by the 4.25 Network and open to the public, the ceremony was held in the Amagasaki Cultural Center and scheduled to start at 6:30 p.m. and continue for two hours. From 5:30 in the afternoon, shuttle buses began departing every ten minutes from Amagasaki Station to deliver attendees to the entrance of the center, where a line of young men and women, dressed in the requisite black attire with an official pin or armband, guided them into the building. When the ceremony commenced at exactly 6:30 p.m., the theater was filled far beyond its six-hundred-person capacity, with people sitting on the steps and standing against the walls. A row of reporters from various media occupied the rear of the auditorium, behind a line of cameras on tripods. During pauses in the program or the songs between the speeches, a steady stream of people continued to enter and exit the event.

The matter of trust emerged in the speeches given by two men who had lost family in the accident. The first speaker, Shinohara Shinichi, was a middle-aged man whose son Takuya had just begun attending classes at Osaka Electro-Communication University fourteen days prior to the accident. His body was found in the second carriage of the derailed train—the carriage that was wrapped in a < shape around the side of the building. The second speaker, Asano Yasokazu, was a slightly older gray-haired man with silver-rimmed glasses, whose wife and sister had been killed in the accident. Asano's daughter had also been severely injured in the accident and was still undergoing physical rehabilitation a year later. Both men delivered heart-wrenching accounts of their losses, while struggling at points to hold back tears and maintain their composure. The audience was clearly moved. Sniffles became increasingly audible throughout the darkened auditorium, and a woman sitting beside me began crying. Such emotional impact was what the ceremony planners had expected, it seemed, as each speech was followed by an invigorating musical performance from amateur and professional musicians and singers that appeared chosen for its potential efficacy in pulling an expectedly empathetic audience back from the tremendous sorrow of the stories. It was also clearly what the media was waiting to capture. Every instance in which Shinohara or Asano began to lose composure prompted a frenzied clicking of camera shutters from the reporters standing in the back row.

Whereas other performances and accounts from accident victims in the ceremony refrained from criticizing JR West, Shinohara and Asano used the opportunity to deliver harsh critiques of the company that focused in particular on its efforts to rekindle trust within the commuter community. It is difficult to say how successful they were in this regard. Although both men appeared to read from a prepared speech, what they each delivered was very much a disjointed stream of consciousness that reflected their distraught emotional states. The result was more affect than argument. Had I not recorded the ceremony, I imagine I might have dismissed both speeches as noble but ultimately unsuccessful. It was only by listening to the recording countless times that I began to see Shinohara's and Asano's speeches as initial articulations of what it would mean to try to understand the accident not as an instance of human error but rather as a provocation to interrogate the nature of trust under techno-social conditions.

Beyond Apology

Standing under a spotlight at center stage and reading from a single A4-sized paper held in shaky hands, Shinohara began his talk with a detailed account of his and his wife's efforts on the day of the accident to determine the fate of their son, followed by various recollections of his son weighed down by expressions of regret over his perceived failures as a father. At some point around halfway through his talk, however, Shinohara shifted to what seemed to be his central and most urgent question: "Why did my son have to die because of JR West's obsession with minutes and seconds?" Underscoring what could only be called the cruel irony of the tragedy, he described his son as someone who, from an early age, had been completely unconcerned about being on time, but who absolutely loved the trains. Offering an image that no doubt resonated with other parents in the audience, he told of taking his son at the ages of two and three years old down to Itami Station every weekend and his son's excitement at just watching the arriving and departing JR West trains. "Who would have imagined that he would die riding that train?" Shinohara lamented. "We never once imagined the need to tell our child to be careful when riding the train and not to ride in the first front cars," he added before declaring that, "JR West has committed a serious crime by betraying our trust and the trust of the community." As if to demonstrate the magnitude of the crime, Shinohara imagined aloud the future that had been taken from him with his son, picturing the woman his son would marry, the people he would help, the child he would raise. The community's trust, Shinohara asserted, could not be recovered simply through formulaic apologies and poster campaigns. It would require something entirely unprecedented: "We want JR West to take drastic actions that are not bound by its regulations and rules. To acknowledge that it has done a terrible thing, an unimaginably terrible thing."[44]

Semantically speaking, an apology encompasses this kind of admission of wrongdoing. But in asking for a response not bound by regulations and rules, it was clear that Shinohara was not seeking merely a better apology, whether that meant more sincere or less formulaic. What he wanted, rather, was something that JR West had thus far entirely failed to articulate. Again and again, JR West's president had knelt down and prostrated himself, pressing his head against the ground and declaring, "We are deeply sorry." But not once had the company asked the accident victims and the community for forgiveness. Japanese has perhaps more than its fair share of words for apologizing: *owabi mōshiagemasu, ayamarimasu, mōshiwake naidesu, shazai shimasu, gomennasai.* In some ways, these terms facilitate what might be called a nonapology apology. They have become the parlance of the nation's weekly scandals and the captions for the accompanying hackneyed images of a company president or state minister prostrating himself before a row of flashing cameras. The term one hears far less often, if at all, is "please forgive us" (*yurushite kudasai*).

This distinction between giving an apology and asking for forgive-

ness is crucial. To ask for forgiveness is an inherent admission of absolute culpability.[45] That is, if one asks for forgiveness, one assumes responsibility for both the act and the intent. An apology does not carry
the same degree of culpability. One can apologize for many things, including things one has no hand in causing. For example, I might say,
"I'm sorry that you are sick." But that does not mean that I am taking
any responsibility for causing your sickness. One apologizes for an act
or a condition, not an intent. If to ask for forgiveness is to assume a
moral culpability, to apologize is to leave oneself room to negotiate degrees of culpability. Apologizing leaves open the possibility that some
cause outside of one's control, some kind of force majeure, might have
brought about the deed for which one is apologizing.

What is more, in asking for forgiveness and thus assuming absolute
and moral responsibility for something, one places one's fate in the
hands of those who have been hurt or wronged. To ask to be forgiven
is to implore an other whom you have wronged to make space for you
again in their world when you perhaps do not deserve it. Such space
does not itself constitute trust. It is rather an emptiness, a placeholder
or gap, that is the condition of possibility for renewing a relationship
from which trust must be gently elicited. Part of the pathos of Shinohara's talk derived from the fact that, insofar as he was expressing a
desire for JR West to ask for forgiveness, he was using the time in front
of the audience to ask his son for forgiveness too. This became clear at
the end, as Shinohara closed his talk by describing his life as having
become "a matter of waiting for death and the opportunity to meet
my son again, to ask forgiveness for creating a society in which a mere
minute-and-a-half delay robbed him of his future."[46]

Forgiveness

Shinohara's talk had an intimate and confessional tone. By contrast,
Asano labored to separate his thinking from his own inner psychological turmoil in order to formulate a critical meditation on questions of
trust and safety. Asano's talk was the final event in the ceremony. It
was supposed to be followed by a choir performance of "Do Not Stand
by My Grave and Cry" to bring the ceremony to a planned end at
8:30 p.m. At 8:50 p.m., however, Asano was still onstage delivering his
reflections. The master of ceremonies hovered at the edge of the stage,
just outside the circle of stage light, looking clearly anxious about

Asano's inattention to the schedule, but also clearly unable to bring herself to interrupt him.

Asano's speech was a noble attempt at a structured critical analysis of safety and trust that unfortunately dissolved into fragmented and incoherent sentences. If the audience seemed sympathetic at first to Asano's plight and willing to strain toward comprehension, as the talk went over time, it clearly became impatient. All around me in the auditorium, people began collecting their things in preparation for the lights to come on and the doors to open. The woman sitting next to me, who was accompanied by her teenage son and who had cried during Shinohara's talk, began putting her jacket on just before half-past eight. She then sat eagerly on the edge of her seat, checking her watch repeatedly and waiting, it seemed, for the opportunity to bolt from the auditorium. To be honest, their impatience was understandable. And yet it was not without irony, as what Asano was asking people to consider was the idea of creating a society with a more forgiving sense of time.

Even with its fragmented and incoherent sentences, Asano's speech was the easier of the two speeches to understand. Asano presented an argument for the pursuit of "absolute safety," by which he meant the total elimination of the possibility of technological accidents. His reasoning embarked from a critique of the "Safety Improvement Plan" and "Safety Verification Plan" that JR West presented in its poster campaign to earn back trust. For Asano, these plans promised nothing more than "relative safety" (*sōtai teki na anzensei*), which he clarified for the audience as meaning "a safety that takes into account a certain unavoidable level of uncertainty and risk." Absolute safety, he asserted, was not something that JR West could provide. The community could not simply entrust itself to JR West. Rather, it was a condition that was realizable only through the active engagement of the community in thinking, practicing, and demanding safety. In accordance with this, Asano seemed to want to transform the ceremony from an event for commemoration and mourning into a space of critical reflection on collective governance of risk or, as he put it, a "forum for public discussion on the true nature of safety." "I want everyone to start thinking about the problem of safety as concerned citizens. We must deal with this as a public issue," he entreated the audience. "No matter what they write about 'Safety improvement plans' or 'reforming their corporate environment,' we the passengers must monitor and supervise, we must enact social supervision to ensure safety."[47] As if to model what this process of "social supervision" might look like, Asano devoted much of his talk to scrutinizing JR West's investigation of the accident. Whereas JR West focused on the question of human error, Asano turned attention to the insufficient *yoyū* in the *daiya*.

In advocating social supervision of risk, Asano to a certain extent echoed Sheila Jasanoff's call for the democratic governance of risk in an essay that attends to the organizational failures that marred local and international responses to the Indian Ocean tsunami of 2004 and Hurricane Katrina in New Orleans in 2005.[48] In her argument, Jasanoff calls for a shift from risk management to risk governance. Whereas risk management assigns risk to the domain of experts and technicians who produce a certain rational calculus of probabilities and acceptable levels of risk, risk governance, as Jasanoff imagines it, would seek to understand and intervene in the sociopolitical environments that produce both risk and disasters. The social governance of risk makes risk a matter of social and political processes of representation. In so doing, it rejects the underlying thesis of Beck's risk society: that risk can only be subdued through forces of reason, whether administered by experts or the science-savvy population of a reflexive modernity.

Where Asano's notion of the social supervision of risk departs significantly from Jasanoff's is in the latter's hesitancy in promising "absolute safety." For Jasanoff, the democratic governance of risk may "foster discovery and innovation" as well as take into account the variant needs of different publics and cultures, but it does not change the fact that "zero risk is an unattainable ideal."[49] Perhaps because he was feeling the growing impatience of the audience at that point, Asano's thinking concerning absolute safety lacked a certain conceptual clarity. His presentation of the concept was made difficult to understand by the fact that he continually tacked back and forth between concern for social, technological, organizational, and psychological conditions without laying out connective logic between those concerns. That Asano was trying to articulate a critical intervention into concepts of risk and safety as a human-machine system would only become clear to me when I revisited my recording of his talk years later, after reading reports from study meetings concerning the accident, meetings in which Asano took part. I will return to that critical intervention into risk and that report in the next section; in the meantime, I want to turn to Asano's closing lines as a moment that not only brought him into dialogue with the notion of forgiveness that emerged in Shinohara's speech but also gestured to the thinking of yet another father of an accident victim who was not at the ceremony.

In concluding his speech, Asano began by thanking the audience for their patience. Offering a final remark, he suggested that only through the realization of absolute safety would the community be able to appease the souls of the "106 deceased." Asano then paused for a moment before reiterating the remark with the revision of "107 deceased," so as to include the twenty-three-year-old driver, Takami Ryūjirō, within the fold of souls to be appeased.[50] This minor amendment carried immense significance. Prior to the anniversary, JR West sent a questionnaire to the accident victims and bereaved families inquiring whether the driver should be counted among the casualties. The majority answered no, and throughout the day the phrase "106 casualties," rather than "107 casualties," was used in all official announcements and ceremonies. Naturally, by offering to exclude Takami, JR West was implicitly allowing blame for the derailment to fall almost exclusively on the shoulders of the novice twenty-three-year-old driver. Members of the community understood this, and they realized the unfairness of it. Without exception, everyone with whom I spoke held Takami only partially responsible for the accident, recognizing that he too was a victim of JR West's abusive practices and its unforgiving system. Yet electing to exclude him from the number of victims suggested that they could not (yet) forgive him. No doubt, Asano's revision to include Takami must have angered some people in the audience. But by accepting the driver among the victims, Asano was gesturing to forgiveness and the willingness to make a space for the man's soul within the community, despite the feeling that he perhaps did not deserve such compassion. In so doing, he inverted the process of asking forgiveness, taking it upon himself to forgive and thus create a space for the emergence of a relationship of trust while urging the community to do the same.

Asano's motion to forgive was not directed at JR West, only the driver. That was not the case, however, for Shike Keijiro, whose son Takashi was a fourth-year university student on the way to a job fair when he was killed in the second car of the train. Shike was a civil servant assigned to a city hall in an area just outside central Osaka. As with Mika, I was introduced to Shike through my in-laws. My father-in-law had worked with Shike at the city hall, and the two had remained in contact. My in-laws and my wife joined me when Shike visited us on a Sunday morning to talk, sitting around the kitchen table as we waited for an order of sushi that my mother-in-law, in her typical pre-emptive mode of consideration, had thought to call several days before. Unfortunately, when the order arrived, everyone seemed to have lost their appetite. For the length of our talk, the gleaming morsels of fish over rice lay untouched on the black lacquer tray with its quintessential gold-inlay flower pattern. Later, as he was leaving, my mother-in-law insisted that Shike take the entire tray home.

Out of deference in part to my in-laws, who had gone to great lengths to arrange our talk, but even more so out of respect for Shike, I did not record the interview. Although I am sure that Shike would not have objected to my recording our talk, I did not feel comfortable asking him. In retrospect, I regret not asking, as my notes from the talk turned out to be highly inadequate. At the same time, I recognize now that I was preoccupied by a sense of guilt. Asking a man who had lost his son to speak with me for the sake of research of which the central concerns were still unclear to me seemed inexorably opportunistic. Thus, the presence of a recorder on the table, no matter how miniaturized the device, could only amplify what felt to me like an intrusion into someone's grief. My concerns, it turned out, were unwarranted, as Shike had found peace in telling his son's story.

Takashi almost never took the Fukuchiyama Line, Shike explained. The job fair was a big deal for him, and in preparation he had purchased a new suit. It was not an expensive one, but rather the kind that most new or prospective employees might wear. His mother adjusted his tie that morning before he rushed off for the train. That was the last time she saw him.

JR West did not permit Shike and his wife to see their son's body after the accident. It was not in a condition to be seen, they told them. Instead, Shike and his wife had to verify their son's remains by means of a Polaroid picture of his feet, and only later through DNA. In place of his body, they received only his possessions, including a briefcase, a *keitai*, and an MP3 player. They also received his new suit, which arrived soaked with his blood. A final image of their son as he boarded the train at a station before the accident was also retrieved from one of the platform security cameras, but Shike and his wife could not bring themselves to watch it for six months. Similarly, their son's suit remained unwashed for months, stained with the blood, a haunting substitute for the body they were never able to see.

Shike and his wife, in contrast to Shinohara and Asano, elected not to speak to the media or participate in events like the public-memorial ceremony. They chose as well not to join the 4.25 Network, nor any of the other support groups. Shike, unlike many of those who were injured in the accident or were families of the victims, had few harsh words for JR West. He said that he did not see the point in criticizing the company, since it would not bring back their son. He and his wife wanted, rather, to be able to begin to forgive JR West. Their desire to forgive did not stem from any theological affiliation or belief; rather, it emanated from their sense of civic responsibility. They wanted, or rather needed, to be able to trust JR West again. To this end, Shike and his wife thus chose to send JR West their son's suit—the one part of him that remained to them. Their gift, so to speak, to JR West was accompanied by the request that it be presented annually at training seminars for JR West employees along with the story of their son, in order to impress on employees the preciousness of the lives they carry in their trains and their responsibility to the community. In return, JR West gave Shike and his wife a video featuring the safety seminar, which Shike played for us that morning. In the video, their son's suit is exhibited on a stand before a classroom of new employees seated at rows of tables as they listen to the story of a young man wearing a new suit who eagerly boarded a train one day to a job interview in the city on the way to his future.[51]

4.25 Network's Study Group

Asano's gesture to forgive the driver and Shike's willingness to forgive JR West were clearly driven by a pragmatic concern for realizing better safety on JR West trains. Such practical sensibility does not diminish the significance of their actions as attempts to create a space for the emergence of trust. If, for JR West, asking for the community's forgiveness would have meant asking the community to make space for JR West in the community again, by taking the initiative to forgive JR West, Asano and Shike were forcing JR West to recognize them as partners in a process of trust as interdependence (*shinrai*).

A similar gesture for creating trust informs a report published in 2011 by the 4.25 Network under the exceptionally long title *Report of the Study Group on the Subject of the Amagasaki Derailment Accident: An Explanation of the Structural and Organizational Problems Surrounding the Accident and a Path toward the Reconstruction of Safety.*[52] The report is the product of an initiative the 4.25 Network undertook five years after the accident and embodies the group's attempt at reconciliation in order create a space of dialogue with JR West as a precondition of trust. The initiative was prompted partly by the revelation, in 2009, of email and phone correspondence between JR West's then-president Yamazaki and members of the Aircraft and Railway Accidents Investigation Commission during the latter's official investigation into the Amagasaki derailment. The revelation undermined the community's faith in the promises Yamazaki had made, upon taking over JR West, to improve safety and undertake serious structural and organizational reform. But overall, members of the 4.25 Network were compelled by the general sense of a lack of genuine closure and persisting questions concerning why the accident happened and why their loved ones had to die.

This separation of what was essentially a single matter into two different questions—why the accident happened, and why their loved ones had to die—reflects the reconciliatory approach that guided the 4.25 Network's interaction with JR West, and it inflects the language throughout the report. As the authors of the report explain, in pursuit of the truth behind the accident and for the prioritization of safety, the 4.25 Network proposed to JR West in September of 2009 to set aside issues of culpability in order to establish a study group that could serve as a place of dialogue between bereaved families and JR West. When JR West agreed to the idea (seven months later) a forum was established with the aim of clarifying organizational and structural problems in JR West, with specific focus on the issues of off-train (re)education training, composition of the *daiya*, fail-safe technology, and JR West's safety-management protocol. In the introduction to the report, the authors make clear the stakes and challenges of the meetings: "This kind of face-to-face dialogue between the company and the bereaved family was the first in the history of accidents in our nation. There was a need for both parties to find a way to proceed with discussion while acknowledging conflicting stances and different opinions in order for both parties to confront the accident squarely. The bereaved family members needed to maintain their calm and JR West needed to demonstrate its sincerity. The display of mutual respect and readiness to logically extract the problem was indispensable."[53] It seems JR West and members of the 4.25 Network kept their promises: at the end of the report, the authors commend JR West for engaging in the forum with sincerity. For their part, the authors clearly take pains throughout the document to present the complexity of JR West's position, from a legal standpoint, during the initial accident investigation while being careful to use blame-free, neutral language, sometimes at the cost of conceptual clarity.

The Human Technological System

One of the main issues the 4.25 Network voiced in their dialogue with JR West concerned the company's safety-management system. The system, they worried, merely conformed to an existing model of risk assessment and therefore only limited rather than eliminated the possibility of another accident. In other words, they feared that the system accepted certain epistemic limits operating within the given paradigm of risk analysis: the "unknown unknowns," or *sōteigai* in Japanese. For the bereaved families in the 4.25 Network, this made the system inadequate. As Asano declared in his speech at the public-memorial ceremony, nothing less than "absolute safety" was acceptable.

Is "absolute safety" realizable? In its attempt to address the problem of safety on the train line, the 4.25 Network's approach remained confined to interventions in the organizational and structural determinants behind the Amagasaki derailment. Its operative premise was that accidents can be "organized away" or designed out of existence.[54] As suggested earlier, the 4.25 Network's strategy toward safety thus shared an affinity with the notion of "democratic risk governance" that Jasanoff advocates, which seeks to intervene at a collective level in the sociopolitical environments that produce both risk and disasters. Shinohara, Asano, and Shike's innovation in this regard was to gesture toward thinking about social governance with the gap. By this I mean that they posited an interdependent form of trust (*shinrai*) as essential for the elimination of accidents. Such trust denotes a relational process elicited within the gap in the space generated through forgiveness—the *yoyū* that one makes to forgive something that is otherwise unforgivable.

Insofar as this kind of thinking with the gap attended to the social and organizational determinants of the accident, it still left unattended questions of the technological apparatus and the so-called epistemic limits of accident prevention, thus failing to address the question of *sōteigai*—"unknown unknowns." In the following section, I turn to one of the most recent reports produced in the wake of the Amagasaki derailment, titled *The Report of the JR West Safety Follow-Up Meetings* (hereafter referred to as the *Follow-Up Report*).[55] The eighty-three-page document is the product of a study group that first convened in mid-2013 and met again on eleven separate occasions over the course of two years. What stands out in the report is the group's attempt to address the question of *sōteigai* through critical reflection on the nature of trust in human and technological relationships. What is more, the group's thinking about the Amagasaki derailment implicitly emerges as a means for thinking critically about the questions of *sōteigai* that arose in the wake of the nuclear meltdown in Fukushima in 2011.

Asano was one of the two representatives from the 4.25 Network who participated in this group. Other participants included three representatives from JR West, three academics, and the former mayor of Amagasaki City. JR West's general director of transport and the head of JR West security also participated in a number of meetings. Similar to the study group initiated by the 4.25 Network in 2009, the *Follow-Up Report* meetings aimed to bring families of accident victims together with JR West in the interest of developing a deeper understanding of the causes behind the accident and its social impact. The results, however, were far less encouraging than for the previous study group. For a large part, confess the authors of the *Follow-Up Report*, the meetings failed to achieve their objective. From the perspective of the bereaved families, JR West did not engage in discussions in any meaningful way, remaining instead entrenched in its position and reiterating conclusions and opinions conveyed in its own investigation.

The *Follow-Up Report* group did not undertake any new empirical analysis of the accident. Rather, it closely scrutinized the methods and conclusions of four primary investigations: the Aircraft and Railway Accidents Investigation Commission's report (2007), the 4.25 Network's report (2011), a proposal from JR West's internal committee investigating the information leaks during the commission's investigation (2011),[56] and JR West's Safety Improvement Plan (2005).[57] The last document is available on the JR West website.

Trusting Machines

When the members of the *Follow-Up Report* group convened in 2012 to discuss the Amagasaki derailment, the ongoing reactor meltdown at the Fukushima Daiichi nuclear-power plant and questions of *sōteigai* were clearly on their minds.[58] The unfolding social and environmental disaster resulting from the meltdown was comparable to the Chernobyl nuclear-power plant accident in Soviet Russia in 1986. As in the Chernobyl accident, communities around the Fukushima plant were forced to evacuate. Many of the surrounding neighborhoods remain abandoned to this day, six years later. But many other families and individuals in areas that were not under forced evacuation also left their homes in fear of the radiation leaking from the ruined reactors. Low-level radiation is expected to continue leaking from the site over the next several decades, while huge quantities of water must be sacrificed daily in order to keep the melted cores from disintegrating further.

As mentioned earlier, the Fukushima Daiichi nuclear-power plant was owned and operated by the Tokyo Electric Power Company, otherwise known as TEPCO, which came under severe criticism in Japan and in the international community for trying to evade both legal and moral accountability for the meltdown. Notably, TEPCO held to the claim that the meltdown was an unforeseeable (*sōteigai*) result of a natural disaster and thus beyond expectations calculable using existing risk-management models and diagnostic technologies.

Nuclear energy was sold to the Japanese population with the promise of absolute safety guaranteed through technical expertise and governmental oversight. One need only look to the television commercials that TEPCO produced in the decades prior to the meltdown to understand how the company engineered the public's trust via claims of absolute safety built on tropes of expert science and infallible rational administration. The meltdown exposed the flimsy ground on which that trust was built, revealing a complex culture of collusion between nuclear regulators and plant operators that had allowed obvious flaws to go uncorrected for years.[59] Emergency backup generators were stored in the facility's basement, where they would be inundated with floodwater; the plant was staffed with poorly trained technicians who had only a minimal grasp of disaster-safety protocol; and warnings about the potential of an unusually big—but not unprecedented—tsunami had been disregarded or buried.

In 2012, the sense of crisis surrounding the nuclear disaster in northeast Japan was still very much at its peak. With radiation levels around the beleaguered power plant still extremely high, TEPCO remained unable—even with robots—to enter the facility and assess the damage. Moreover, large amounts of contaminated cooling water collected in gigantic, hastily constructed steel tanks were leaking into the ground and sea, while doctors began noticing significant increases in thyroid irregularities in children from the area—information that governmental bodies did their best to disavow and or dismiss. Although the name *Fukushima* and the term *nuclear disaster* never appear in the *Follow-Up Report*, the Amagasaki derailment emerges from the report as an analytic through which to engage the question of *sōteigai* in the Fukushima nuclear accident. In particular, toward the end of the report, the authors define technology in the context of reflections on "humans and technology" (*hito to gijutsu*) in what reads as an attempt to develop a guideline for technological development that recognizes the limits of human knowledge and control:

Technology, broadly defined, is an artificial means to introduce natural power (including "hard" forces such as energy and material, and "soft" forces such as the law of nature, natural phenomena, inherent properties in nature) in order to achieve a specific objective efficiently and safely. Insofar as technology is an [artificial means] to introduce natural power, technology itself operates according to the laws of nature. It will thus not always conform with the desire and expectations of man and society. In this sense, technology possesses powerful natural properties. No matter how sophisticated its automation, technology is conceived and produced by humans and cannot operate alone. It is up to society and companies to determine what technology to develop, how it is to be used, and what objectives it is to achieve. In this regard, technology possesses powerful social properties (Ishitani 1972). Human beings are involved in [*kakawatte ori*] the manner in which technology is created and used, thus the way in which technology is developed and used is strongly informed by the intentions of those involved [*hito no ishi ga tsuyoku eikyō shite iru*]. For this reason, in looking at the behavior of technology, it is imperative to maintain a point of view that acknowledges technology as a human-technological system [*hito-gijutsu shisutemu*].[60]

Embarking from a fairly standard and instrumental understanding of technology, the definition at first hearkens back to the nineteenth-century German theorist Franz Reuleaux's seminal delineation of the machine as a means for deploying the "mechanical forces of nature" toward the realization of human goals.[61] Similar to Reuleaux's theorization of machines, technology for the authors of the *Follow-Up Report* is irreducible to a binary distinction between nature and culture. Technology is a human fabrication and thus an expression of culture, yet as a medium of natural force, it is subject to laws of nature that are external to the dictates of human beings. Straddling nature and culture, technology takes on an ontological indeterminacy that has one critical ramification—we cannot expect to maintain complete control over machines. Insofar as machines are products of human design, their operation according to the laws of nature renders them quasi-natural objects and thus they will not always comply with human expectations and demands. The definition continues, specifying that technology is not an autonomous force— it "cannot operate alone"—meaning that it is inseparable from human society. Attributing to technology a certain degree of agency, the definition recalls Latour's formulation of Actor Network Theory (ANT).[62] At the same time, the authors go beyond Latour's exposition of human and machine processual interaction by placing the specific burden of realizing a techno-ethics on human beings: "It is up to society and companies to determine what technology to develop, how it is to be used, and what objectives it is to achieve." Taken in context with the preceding claims, we are asked to read this assertion as referring to far more than technology's instrumental value. It implies the idea that human beings bear the responsibility for discerning between negative and positive technologies, between technology that threatens and constrains life and technology with which we can foster a mutually beneficial and ethically coherent interdependence. This idea is reinforced and complicated in the final lines, which relegate humans to participants in, rather than masters over, technological development. Humans are only *involved in* the creation of technology. Insofar as human intentions may *strongly inform* technology's development and use, they do not determine it. The formulation is sobering. It undercuts centuries of human conceit founded on the claim to occupy the position of homo faber, controlling life and environment through a unique capacity for technological innovation. What is more, the formulation is pivotal, as it emphasizes human and machine as bound together in a relation of interdependence. Machines are not things, objects, or instruments. They are partners, so to speak, composing a "human-technological system": a collective. Sidelining debates over epistemic, regulatory, or operational constraints, the formulation acknowledges technology as *an other* within a system of ontologically entangled relations that can best be described as ecological. Consequently, "human error" becomes a non sequitur. There can be no such thing. Error of any kind is always a matter of human-machine interplay. Refusing to bestow ontological priority on either human or machine, the formulation demands instead that we look to the dimension of interplay—the margin of indeterminacy—to understand and circumvent disaster.

In taking this approach, the authors gesture toward a de-totalization of the term *technology* in ways that evoke the notion of technicity. Refuting an understanding of technology as either purely instrumental and value-free or subject to the dictates of human will, they ask to understand and discriminate between machines in terms of the quality of collective they enable. What emerges from the report is thus not just a criticism of JR West but rather an attempt to articulate a much broader critique of the human-machine relationship that has characterized postwar Japan. The authors go on to add two principles or guidelines for technology in light of their reflections on the nature of human and machine interaction:

1. Deployment of any machine must take into account the limits of the material properties and machine performance.
2. No machine should be deployed that can potentially become unresponsive in catastrophic conditions.[63]

With these principles, the authors move from the ontological questions *What is technology?* and *What are its relations to human beings and society?* to the matter of what machines can be trusted. In this regard, the second principle takes on the greatest gravity and, as the authors note, is already implicit in principle 1. In specifying that a machine should not be used if it has the possibility of becoming unresponsive under catastrophic conditions, what they mean, essentially, is that machines must always operate with *yoyū* so as to be able to maintain a margin of indeterminacy. This is not simply about technology failing safely. Rather, it concerns the machine's capacity to remain in a trusting relationship with its human operators. A machine that can maintain a gap in order to remain in relationship with its human operators despite catastrophic conditions is a forgiving machine and thus a trustworthy machine. Trust, according to this formulation, is no longer confined to specifying merely a psychological facet in the relationship between humans and machines. Rather, it takes on the quality of a material force that enables a coherent collective unity. Trust becomes the condition of possibility of that collective unity. It is that which allows for generative interdependence through communication across systems (or milieus) with different orders of magnitude. Such a notion of trust demands cultivating a different kind of relationship with technology as well as a different kind of technology.

Conclusion

Since 2010, I have returned to Hyogo Prefecture every summer. Each trip has given me countless opportunities to ride the train on the Fukuchiyama Line. The dilapidated metal shops and rusting warehouses that once stood alongside the tracks between the Itami and Amagasaki stations have given way in many places to rows of prefab homes and small shopping centers. The apartment building that the train slammed into is finally being torn down after being vacant for years. Few people in the train car seem to look up anymore from their digital devices or newspapers as the train goes around the curve. Nevertheless, the Amagasaki derailment has not faded from memory. JR West and members of the community have continued to gather annually at the accident site to commemorate the victims and to mark the time of the tragedy, and a permanent memorial has been constructed next to the building. In addition, on the recent twelve-year anniversary of the accident, NHK, Japan's public broadcasting service, featured a special segment that showed JR West drivers gathering each morning before embarking on their daily duties in order to recite the company pledge to remember the Amagasaki derailment and the lives that were lost. Friends and acquaintances from the community with whom I have spoken each year concede that JR West has done much to regain their trust and that they feel safer riding the train. No matter the extent of delays, the train unfailingly slows to sixty kilometers per hour or less when rounding the curve before Amagasaki Station.

The Amagasaki accident did not spur the development of a new kind of technology. Aside from a few changes to safety systems and suspension, the commuter trains running on the Fukuchiyama Line are more or less the same as the train that derailed in 2005. What the Amagasaki accident did was open a space of critical intervention that allowed for the development of a different relationship with JR West and a reconceptualization of the commuter train as technology. As I have tried to make clear, this was not at all an easy nor instantaneous process. Its difficulty and duration were due in part to JR West's intransigence regarding the position and role it perceived for itself within the commuter community. At the same time, it was no less a result of the community's struggle to make sense of the accident and its inchoate understanding of its own responsibility for the emergence of the conditions behind the accident. The difference between JR West and the community lies in the disparate orientations they took in treating the ninety-second gap that led to the derailment. Whereas JR West treated it as a performance failure, for the community the gap became a medium of intervention through which to reflect on the significance of its relationship with JR West and the commuter train.

What is more, the gap became an analogy for the reconceptualization of trust as an unfinished process fundamental to the creation and maintenance of a relationship between heterogeneous collective realities, whether between a community and a corporation or between a train system and a commuter population. In the discussion that emerged from the Amagasaki derailment, trust became irreducible to a thing or property within a relationship. It became a process contingent on maintaining a certain space or gap of interaction and accommodation. Trust was thus transformed from something an institution, person, or even technology possesses into a quality that must be constantly generated to maintain the energetic interactions of a collective's margin of indeterminacy. Where the Amagasaki accident differs from the many train accidents in Japan's history that preceded it is in the success it has had in keeping the gap open as a space of intervention for over a decade.

By virtue of the gap it held open, the Amagasaki derailment became a means through which to think about questions concerning technology, corporate responsibility, and trust that reemerged as a result of the reactor meltdowns at TEPCO's Fukushima Daiichi nuclear-power plant. As I have argued, the attempt among the authors of the *Follow-Up Report* to (re)define the relationship between humans and machines in respect to the problems of trust and the seemingly ineluctable nature of technological accidents that were raised by the Amagasaki derailment was clearly inflected by the emerging uncertainty and anxiety around the Fukushima nuclear meltdown. Although the authors never make the connection explicit, the guidelines they lay out for a trusting relationship with technology and for ethically responsible technological development can be read as a direct criticism of TEPCO's operation of its power plant as well as of nuclear power itself as an inherently flawed "human-technological system."

Notes

1. The official number of victims in the Amagasaki derailment was 107 dead and 562 wounded. See Japan's Aircraft and Railway Accidents Investigation Commission, *Tetsudō jiko chōsa hōkoku sho*, 1.
2. Haiyama Kikuo, interview by the author, July 3, 2005. I spoke with Haiyama outside his shop three months after the accident when I was visiting the accident site.
3. "Hijō burēki de sharin rokku ka kenkei ga jiko to kanren chōsa," *Asahi Shinbun*, April 28, 2005.
4. A few days after the initial scandal broke, newspapers carried an additional story of a veteran driver who had committed suicide in 2001 after being subjected to three days of day-shift (re)education for a fifty-second delay incurred when checking a safety device.
5. "Kyū nin ijō shibō no tetsudō jiko, 71 nen no kintetsu tokkyū irai" [Train accidents killing more than ten people until the 1971 Kentetsu Limited Express], *Asahi Shinbun*, April 25, 2005.
6. See Bond, "Governing Disaster"; Fortun, *Advocacy after Bhopal*; George, *Minamata*.

7. Jasanoff, *Learning from Disaster*.
8. Giddens, *The Consequences of Modernity*.
9. See Knight, *Risk, Uncertainty and Profit*.
10. Beck, *Risk Society*.
11. A leading thinker in this field in Japan is Hatamura Yōtarō. See Hatamura, *Sōteigai o sōteiseyo!*
12. Beck, *Risk Society*, 3.
13. Giddens, *The Consequences of Modernity*, 26–36.
14. Downer, "When the Chick Hits the Fan," 7–26; Downer, "Trust and Technology"; Downer, "'737-Cabriolet,'" 725–62.
15. Downer, "Trust and Technology," 100.
16. Jasanoff, "Beyond Calculation," 36.
17. Downer, "'737–Cabriolet,'" 726.
18. Downer, "'737–Cabriolet,'"; Downer, "Trust and Technology."
19. Perrow, *The Next Catastrophe*; Perrow, *Normal Accidents*.
20. Downer, "'737–Cabriolet,'" 741.
21. Fujii, "Intimate Alienation."
22. See Robertson, *Takarazuka Sexual Politics and Popular Culture in Modern Japan*.
23. Kawashima, *Naze fukuchiyama sen dassen jiko wa okotta no ka*, 60–61.
24. Ibid., 51–52.
25. In *Naze fukuchiyama sen dassen jiko wa okotta no ka*, Kawashima Ryōzō provides an exhaustive explanation of the incremental improvements made toward the development of lighter, faster, and more comfortable trains on the Fukuchiyama Line and the subsequent increase in ridership. He emphasizes how difficult it still was for JR West to attract riders from the competing Hankyū Line, with commuters choosing sometimes to switch back to the Hankyū lines at relevant points in their commute because of the better service. Hiromi Suzuki and Tesuo Yamaguchi give similar explanations and examples in *JR nishi nihon no tazai: Hattori untenshi jisatsu jiken to amagasaki dassen jiko* [JR West's great crime: the suicide of the driver Hattori and the Amagasaki derailment].
26. Similar explanations and examples are also given in Suzuki and Yamaguchi, *JR nishi nihon no tazai*.
27. In his book concerning the science of timetables, the railroad-information and -system specialist and researcher Tomii Norio cites the *daiya* on the Fukuchiyama Line as an example of a nearly perfectly calculated system. See Tomii, *Resshya daiya no himitsu*, 37.
28. Kawashima, *Naze fukuchiyama sen dassen jiko wa okotta no ka*, 108–9.
29. Reference to this announcement appears in Kawashima, *Naze fukuchiyama sen dassen jiko wa okotta no ka*; see also Noriko Yamane and Koji Togo, "JR nishi nihon no 'tsumi to batsu'" [JR West's "Crime and Punishment"], *Sandei mai nichi*, May 22, 2005, 140–43.
30. Evidence of this emerged in the accident report published by the Ministry

of Land, Infrastructure, and Transport. The report details how drivers were routinely forced to exceed speed limits on all sections of the track.

31. Japan's Aircraft and Railway Accidents Investigation Commission, *Tetsudō jiko chōsa hōkoku sho.*

32. Brockman, *Twisted Rails, Sunken Ships,* 13–14.

33. This was a standing accident-investigation panel established in 2001 at the behest of the families of forty-two passengers killed in 1991 when a JR West train collided with a local train in Shigaraki (Shiga Prefecture). In 2008, the commission was expanded to include the Marine Accident Inquiry Agency and was renamed the Japan Transport Safety Board under the Ministry of Land, Infrastructure, Transport, and Tourism (MILT).

34. See Japan's Aircraft and Railway Accidents Investigation Commission, *Tetsudō jiko chōsa hōkoku sho,* 243.

35. Japan's Aircraft and Railway Accidents Investigation Commission, *Tetsudō jiko chōsa hōkoku sho,* 5; JR nishi nihon anzen forōappu kaigi, *JR nishi nihon anzen forōappu kaigi hōkokusho,* 54.

36. Satō, *Dare mo kataritagaranai.*

37. Mika, interview by the author, June 27, 2005, at Hankyū Kawanishi Station.

38. "Fukuchiyama-sen jiko 1 nen tsuitō reishiki ni 1886 nin sanretsu" [One year memorial ceremony for the Fukuchiyama line accident, a line of 1886 participants], *Yomiuri Shinbun,* April 25, 2006.

39. Japan's Aircraft and Railway Accidents Investigation Commission, *Tetsudō jiko chōsa hōkoku sho.*

40. "Dassen jiko no JR takarazuka sen, 55-nichiburi saikai shihatsu shachō, izokura mokutō," *Asahi Shinbun,* June 19, 2005. All of the newspapers carried more or less the same story.

41. Mika, interview.

42. The series eventually reached volume 83, although at the one-year anniversary of the accident, only volumes 1 and 2 had been published. West Japan Railway Company, www.westjr.co.jp/company/action/poster/.

43. Yamagishi, *Shinrai no kōzō,* 37–42. Yamagishi posits that the term *shinrai* rests on a recognized precondition of danger or uncertainty.

44. Speech by Shinohara Shinichi, Amagasaki Cultural Center, April 25, 2006. My translation, reproduced from my recording.

45. I am in debt to my brother Judge Daniel Fisch for helping me think through the legal implications of this difference.

46. Speech by Shinohara Shinichi.

47. Speech by Asano Yasokazu, Amagasaki Cultural Center, April 25, 2006. My translation, reproduced from my recording.

48. Jasanoff, "Beyond Calculation," 36–37.

49. Ibid., 23.

50. Speech by Asano Yasokazu.

51. Shike Keijiro, interview by the author, January 5, 2007, Hibari-Gaoka, Takarazuka City.

52. 4.25 Network and Nishinihon ryokaku tetsudō kabushikigaisha, *Fukuchiyama-sen ressha dassen jiko no kadai kentō kai hōkoku.*

53. 4.25 Network and Nishinihon ryokaku tetsudō kabushikigaisha, *Fukuchiyama-sen ressha dassen jiko no kadai kentō kai hōkoku,* 1–2.

54. Downer, "'737–Cabriolet.'"

55. JR nishi nihon anzen forōappu kaigi, *JR nishi nihon anzen forōappu kaigi hōkokusho.*

56. See *JR nishi nihon fukuchiyama sen jiko chōsa ni kakawaru fushōji mondai no kenshō to jiko chōsa shisutemu no kaikaku ni kansuru teigen,* published April 15, 2011, accessed October 15, 2017, https://www.mlit.go.jp/jtsb/fukuchiyama/kensyou/fu04-finalreport.html.

57. See *Anzensei kōjō keikaku,* published May 31, 2005, accessed October 15, 2017, https://www.westjr.co.jp/safety/fukuchiyama/plan_improvement/pdf/keikaku_00.pdf.

58. Although it would be years before TEPCO released the data it collected in the initial hours of the disaster, the power outage caused a nuclear meltdown in three of the plant's four reactor cores. See Reiji Yoshida, "TEPCO admits it should have declared meltdowns at Fukushima plant much earlier," *Japan Times,* February 24, 2016.

59. Kainuma, *Fukushima ron.*

60. *JR nishi nihon anzen forōappu kaigi hōkokusho,* 57. My translation. The text referenced in this passage of the report is Ishitani, *Kōgaku gairon.*

61. Reuleaux, *The Kinematics of Machinery,* 33. See also Atsuro Morita's excellent analysis of Reuleaux's machine theory in "Rethinking Technics and the Human."

62. Latour, *Reassembling the Social.*

63. *JR nishi nihon anzen forōappu kaigi hōkokusho,* 58.

Conclusion: Reflections on the Gap

Here we are knee-deep in garbage, firing rockets at the moon.
PETE SEEGER, FOLK SINGER (1919–2014)

Pete Seeger's pithy tune lends itself to at least two different interpretations. On the one hand, it can be understood as an unequivocal condemnation of technology and its corollary modernizing effects. Garbage and rockets are part of the same problem, in this reading. They are the result of a process of technological development that has destroyed the planet, leaving us buried in refuse with nothing left to do but set our sights on the stars and our next target for extractive industry. Such a reading corresponds with a general and popular anti-technology discourse that sets up a false and reductive opposition between the human and the technological, nature and machine. There is no way to go forward with technology, it declares. Our only hope is to recover a lost relationship with nature through traditional forms of indigenous knowledge that allow for living in harmony with the natural environment.

On the other hand, an alternative and more complex reading of Seeger's words would understand the relationship between garbage and rockets as antithetical and ironic. According to this interpretation, the song is asking how it is possible that human society can realize the technology for putting a human being on the moon yet cannot maintain a healthy environment for living organisms here on Earth. The problem is not technology, this suggests, but rather a more fundamental failure of human beings and human society to realize its potential with—not through—technology. If we follow this reading, Seeger is not condemning the catastrophe of technological modernity: he is pointing to the tragedy of our relationship with technology.

This alternate reading of Seeger's song resonates with the strategy I have pursued in developing a technography of Tokyo's commuter train network. My aim has been to think with Tokyo's commuter train network as a collective constituted in the interplay of humans and machines, and in so doing develop a critical exploration of its particular practices and processes while providing a novel anthropological approach to collective life that I call machine theory. This has involved resisting a familiar story of the train as a medium of mechanistic conditioning under capitalist modernity's hyper-rationalizing logic. Machine theory is theory from the gap that offers a critical approach to machines and large-scale technological infrastructures. It is not meant to provide a set theoretical framework for engaging machines and technical ensembles; rather, it encourages a fundamentally experimental and ontologically driven approach toward technologically informed collectives that embarks from thinking with the immaterial and the material forces articulated within a technical ensemble's margin of indeterminacy. Machine theory thus does not specify how to think. It specifies only that one situate oneself within the margin of indeterminacy of a machine or technical ensemble and think critically with the intensities and relationships that emerge therein.

Simondon's philosophy of technics has been my primary inspiration in developing an anthropology of the machine. Of course, Simondon was not an anthropologist. His training was in philosophy. He also held a degree in psychology and studied the physical and engineering sciences.[1] His central intellectual concern was in understanding and exploring the interlaced processes of physical, organic, technological, and psychosocial individuation. His principal question in this regard was, *How do things, organisms, technology, and collective emerge, and what can they become?* While highly critical of capitalism and its influence on technological development, Simondon maintained a generally optimistic view of humanity's potential for collective individuation through technical thinking. Developing Simondon's thinking for an anthropology of the machine concerning Tokyo's commuter train network has required finessing his concepts to a certain degree, which I have done by reading his work in conjunction with a number of thinkers who have taken up his approach toward developing critical interventions into technical ensembles. I have taken what some may think is considerable liberty in this context in foregrounding and elaborating what I believe is an inherent framework for an anthropological exploration of techno-ethics in Simondon's approach to collective. What Simondon offers, I have argued, is an analytical orientation that focuses on the margin of indeterminacy as the dimension of collective constituted through human and machine ontological entanglement.

I can anticipate the critique, spurred by recent literature in anthropology advocating an ontological turn, that by drawing on Simondon I am imposing nonindigenous theory in my exploration of Tokyo's commuter train network.[2] My initial response to this is to emphasize that my turn to Simondon is informed by my encounter with the gap between the principal and actual *daiya*, not the other way around. In other words, my method and analytical approach derive from the material conditions of my site. While I was guided by my training as an anthropologist, my approach conforms as well to the methodological prescriptions of the ontological turn. Furthermore, while I do not disregard the significant contributions that the ontological turn has made in anthropology, I would argue that insisting on a locally informed theoretical engagement with questions surrounding technology in Japan risks ignoring generations of scholars who have labored to departicularize Japan so as to remove it from the shadows of "the Oriental."[3] As these scholars have compellingly shown, in Japan philosophy and social theory concerning technology and culture have been deeply enmeshed in a discourse with thinkers in Europe and the United States since the late nineteenth century, if not earlier, making it impossible to sustain a clear division between (so-called) Japanese and Western (European and American) systems of thought.[4] In fact, scholarship in Japan has produced profound critical explorations of modern society in dialogue with thinkers such as G. W. F. Hegel, Emmanuel Kant, Martin Heidegger, and Karl Marx. Even a nonacademic figure such as the director Ichikawa Kon (chapter 4) not only read thinkers like Max Weber but also saw himself as elaborating his own intervention into Weber's critique of technological rationalization for an audience already familiar with the argument. That this intellectual and philosophical dialogue continues today is evidenced by my discussion with the engineer responsible for developing distributed-autonomy infrastructure (chapter 3). As with much of the discussion emanating from Europe and the United States, scholarship in Japan has tended to treat questions of technology within the framework of an overarching concern with modernity and its effects on culture.

Thinking from within the margin of indeterminacy liberates the critical exploration of technology from the discursive contours of technological narratives of modernity. The latter tends to take the form of a technological history that follows the invention of various technological tools, machines, and/or ensembles and their subsequent feedback effect in shaping human society and human experience. In so doing, it veers toward a technological determinism in which the focus on identifying technologically informed shifts becomes a way of delineating humanity's passage from premodern, to modern, to postmodern periods.[5] At the same time, a technological history inevitably grants humanity ontological priority as the historical subject who fashions the technological object through the application of rational scientific thought, thus setting the stage for existential anxiety and concern over the loss of human agency as the question shifts to how technology shapes society and how technical objects create technological subjects.[6] By contrast, the approach (inspired by Simondon) that I have followed in this book of thinking from within the margin of indeterminacy traces the emergence of a technology through the genesis of its schema of operation. It focuses not on the relationship between humans and machines but rather on the processual relations from which humans and machines emerge within provisionally stable patterns of interaction. The emphasis is on the scene of interplay between human and machine processes in which sedimented patterns of interaction emerge, not on human subjects and technological objects. There is no ontologically pure human subject from which to begin, but rather an entity that is human only through its always-/already-existing ontological entanglement with the material processes of its environment. When thinking from within the margin of indeterminacy, it makes no difference whether a technology is industrial or postindustrial, mechanical or informational. Industrial and postindustrial, modern and postmodern are simply period qualifiers complicit with a modern teleological narrative of technological progress as the fashioning of endless novelty for the sake of driving economic consumption. What matters, rather, is the technicity of a machine—the degree of collective potential that it elicits. Approaching technology in terms of its technicity places emphasis on the quality of collective a technology affords, where quality refers to the degree of a machine's or technical ensemble's margin of indeterminacy and its corollary capacity for a reticular flourishing.

Technicity, I have argued, carries an inherent consideration of techno-ethics. Although Simondon does not put it in such terms, his thinking around technicity allows us to gauge the ethical nature of a technology as directly proportional to the degree to which it embodies an organic schema of relational becoming—what he calls "becoming organic." A technology with a high degree of technicity affords a high degree of collective individuation because the process of concrescence holds the potential for continuing individuation as a reticular process of becoming.[7] In pursuing the question of techno-ethics through technicity, I extrapolated the idea of *yoyū* in a machine or technical ensemble to encompass the notion of forgiveness. A technology with *yoyū*, I argued, is a technology with an expansive and inclusive margin of indeterminacy. Emphasizing the degree of a technology's forgiveness shifts the discussion around the problem of technological accidents from being about thinkable versus unthinkable accidents to being about the question of trust. In contrast with risk-management discourse, in which trust is subsumed under rational thought and organization, trust, when understood through the lens of forgiveness, becomes a matter of thinking with technology toward collective becoming. Trust, in this regard, takes on the quality of a material force that traverses and binds the coherent unity of a technical ensemble.

This book embarked with the claim that operation beyond capacity in Tokyo's commuter train network is a highly localized and particular iteration of a human and machine relation that provides a mode of critical intervention into general formations of collective life. Operation beyond capacity, in this regard, is not presented as some kind of ideal paradigm of techno-social organization. Rather, it is raised as a collective condition that is good to think with. Its significance lies in part in its illumination of the work of the margin of indeterminacy as a dimension of intense interaction and accommodation between humans and machines in the formation of collective life. This underscores the commuter as aggressively attuned to the ambient fluctuations of the network's rhythm rather than as a compliant automaton, but it also brings forward a sobering realization: namely, that operation beyond capacity is not a technologically imposed condition. It is a reality constituted in the cooperative and invested interaction of humans and machines. On the one hand, operation beyond capacity thus reflects a unique human capacity for becoming with technology that is not about the fusion of body and machine to form a cyborg, but rather about the capacity for body and machine to affect and be affected by each other toward novel collective articulations. On the other hand, operation beyond capacity points to a human capacity to learn to inhabit comfortably collective conditions of its own making that are both unsustainable and unendurable. Operation beyond capacity thus comes to embody a potential danger.

The evolution of operation beyond capacity in Tokyo's commuter train network has produced a technical ensemble with a remarkably expansive margin of indeterminacy. With the development in the 1990s of a traffic-control system inspired by an understanding of the principles of emergence in natural systems, the network's degree of technicity ostensibly increased. As I have emphasized, however, its expansive margin of indeterminacy does not necessarily translate into a high degree of technicity. One would be hard pressed to call Tokyo's commuter train network a highly ethical technological ensemble. Indeed, with its emphasis on extreme capitalism, its principle of accommodating the body on the tracks without disruption of service, its crushing packed trains, and the gendering of train cars as spaces of male fantasy in ways that encourage violence against female commuters, Tokyo's commuter train network harbors countless ethical shortcomings. Nevertheless, the aim of this book has not been to hold Tokyo's commuter train network up as a model of a technical ensemble worthy of emulation, but rather to think with it toward a new understanding of collective.

That my thinking in this book ends with reference to the ongoing nuclear crisis around the Fukushima nuclear-power plant is intended to mark the danger we face in failing to reconceptualize our understanding of collective. Since the triple disasters of March 2011, or 3.11, scholars have struggled to find meaning in the tragedy through critical interventions designed to illuminate the lessons from Fukushima. I have argued in this book that the lessons from Amagasaki in many ways prefigure the lessons from Fukushima. On a technicity spectrum, nuclear power registers on the negative end. A nuclear reactor is an absolutely unforgiving machine. In the initial years following the 3.11 disaster, this message seemed to be understood by a great many people in Japan, sending them out into the streets in unprecedented numbers to demand the nation scrap its nuclear program. In recent years, however, the will for change that energized those protests seems to have fizzled, while the Japanese government has embarked on a massive project to restore a former order through the reconstruction of conceptually flawed technical infrastructures. Part of what led the government to pursue reconstruction projects despite objections from environmental groups and concerned citizens, I would argue, has been the failure of the latter to advance a counter theory of technology adequate to imagining a different technological environment. Too often, the rhetoric of environmentalism encompasses a rejection of technology accompanied by nostalgic longing for an imagined lost relationship with nature. As I suggested in the first pages of this book, there is no going back. Rather, we need to transform our understanding of technology if we hope for collective life to not only survive but thrive on this planet. If we can begin by reconceptualizing a commuter train network free of determinations imposed by received frameworks of understanding, perhaps there is still time to make things right.

Notes

1. Nathalie Simondon, "Gilbert Simondon Biography," http://philosophyofinformationandcommunication.wordpress.com/2013/06/30/gilbert-simondon-biography/ (site discontinued).
2. For an excellent summary of the "ontological turn," see Jensen, "New Ontologies?"
3. Since the scholars who contributed to this effort are many, I list here only those with whom I studied and whose work had a direct impact on my thinking for this book: Harry Harootunian, Naoki Sakai, Victor Koschmann, Carol Gluck, Marilyn Ivy, Thomas LaMarre, and Tom Looser.
4. Many of Japan's notable interwar and wartime philosophers sought to separate Eastern and Western thinking in a project they called "overcoming modernity." Ironically, as Harry Harootunian famously argues, in so doing they demonstrated instead the futility of their endeavor since there is perhaps nothing more indicative of a sensibility informed by conditions of modernity than the desire to overcome modernity. See Harootunian, *Overcome by Modernity*.
5. My thinking here is deeply informed by Thomas LaMarre's critique of what he calls a "modernity thesis." LaMarre, *The Anime Machine*, xxiii.
6. LaMarre, "Afterword: Humans and Machines."
7. See Muriel Combes's explanation of reticular becoming in *Gilbert Simondon and the Philosophy of the Transindividual*, 65–70.

Acknowledgments

It takes a collective to produce a book. This book is the product of an especially large collective of people, technologies, and institutions that have supported me for over a decade, from fieldwork through revisions. The initial idea for this book emerged on a late summer evening in Tokyo in 2003, when I missed the last train of the night from Shinjuku Station to where I was staying on the western side of the city. At the time I was a graduate student in the Department of Anthropology at Columbia University, doing preliminary dissertation fieldwork. With nothing better to do to fill the gap in time between the last train in the evening and the first train in the morning, I spent the in-between hours walking around the station, thinking and talking with other stranded commuters. Learning to inhabit that gap in time was the first step in the long journey to produce this book.

I owe special thanks to my mentors at Columbia University for inspiring me in the first iteration of that journey. Marilyn Ivy was my guide and ideal. Her powerful intellect, compassion, and ethnographic sensitivity helped me shape that experience in Tokyo in an anthropological direction. She has remained the voice in my head throughout the years of revisions, urging me always to read a scene more closely and dig deeper with my questions. I thank as well John Pemberton for his constant support and his critical interventions into technology and technological discourse. Without those insights, I would have never imagined that a commuter train system could become a topic of anthropological inquiry. I also owe a deep debt to Rosalind C. Morris, whose approach to media has been a major inspiration in my thought and writing. Many thanks are due to Thomas Looser, whose comments on the first version of this text guided my thinking around media and technology. I owe a very special debt to Thomas LaMarre, who set me on my academic career years ago when I was just starting my bachelor's degree at McGill University and whose curious question—"Can the train teach us to cry?"—at my dissertation defense left me pondering for over a decade about what this book wanted to be. His work has been a major inspiration and deeply informs this book.

The initial fieldwork for this book was carried out between 2004 and 2006 through generous funding from the Donald Keene Center's Shincho Graduate Fellowship for Study in Japan. Without that support, my experience of missing the train home from Shinjuku Station might have simply vanished into the gap. During that research, I was affiliated with Sophia University in Tokyo, where I received excellent guidance and assistance from David Slater. Like so many of the Japan-related scholars of my generation, I am especially grateful for the fieldwork workshops that David organized, which gave me the chance to present my work in progress and to engage with my future colleagues.

In 2009, I was fortunate to receive the Edwin O. Reischauer Institute Postdoctoral Fellowship in Japanese Studies. The year with the institute was extremely productive and academically stimulating, and I am grateful for the opportunity it gave me to share my work with Theodore Bestor, Ian Condry, and Tomiko Yoda.

In Japan I am forever grateful to the many people who took time out of their day, their commute, and their weekends to talk with me and share their stories and sentiments regarding a commuter's life. I owe special thanks to Tomii Norio at the Chiba Institute of Technology for his patience in sitting with me on numerous occasions to explain the minutia of commuter-train operation, for arranging my tour of the train command center, and for providing me with a wealth of diagrams and technical literature. My thanks are due as well to Mori Kinji and to Mito Yuko for the time they devoted to meeting with me and explaining their work.

Many people have helped me over the years by reading and commenting on chapter drafts and pushing me to refine my ideas and approaches. I am especially thankful in this regard for the support I received from Orit Halpern and Junko Kitanaka, who each read an initial draft of this book in its entirety and provided invaluable feedback as well as unwavering encouragement over countless coffees, dinners, and drinks. I would also like to thank the participants of an itinerant infrastructure workshop that I co-organized with my colleague Julie Chu over the years: many thanks to Amahl Bishara, Eleana Kim, Brian Larkin, Andrew Matthews, Jun Mizukawa, and Bettina Stoetzer for their assistance in developing the ideas for this book. This book would also not have been possible without the contributions of time and intellectual input from my fellow machine thinker Morita Atsuro and the media scholar Marc Steinberg. I am eternally grateful to both. I also thank Casper Jensen for reading and commenting on early chapter drafts. His unusual ability for parsing complex theory in graspable terms proved enormously helpful.

This book realized its final shape within the intellectually enriching milieu of the Department of Anthropology at the University of Chicago. When I began teaching in the department in 2010 my ideas for an anthropology of the machine were still very much in nascent form. The University of Chicago gave me the opportunity and the space to develop those thoughts, and I owe my deepest gratitude to all my colleagues there for their support, intellectual conversation, and friendship. Thanks are due as well to Anne Ch'ien, Sandra Hagen, Kim Schafer, and Katherine Hamaguchi for their daily assistance in countless administrative matters. I am especially in debt to Joseph Masco and Kaushik Sunder Rajan for reading and commenting on early versions of the first chapters, to Judith Farquhar for guiding me through the first years, and to Julie Chu for her encouragement and ethnographic insights. Also, without Kathy Morrison's finesse in negotiating a year of sabbatical, I would not have been able to complete this book.

I developed many of the ideas for this book by teaching a graduate seminar in machine theory at the University of Chicago. In 2013, I benefited greatly from co-teaching that seminar with Poornima Paidipaty. I also owe numerous insights to the many excellent graduate students who participated in that seminar and endured my conceptual experimentation and explorations with great patience. Special thanks are due to Love Kindstrand and Hiroko Kumaki for reading many works on Japan, technology, and nature with me.

I have also benefited greatly from the support of Michael Bourdaghs, James Ketelaar, Chelsea Foxwell, and Hoyt Long, my colleagues at the Center for East Asian Studies. Since 2010 I have had the great fortune of pursuing follow-up research for this book through generous funding provided by the Lichtstern Endowment for the Department of Anthropology and the Center for East Asian Studies at the University of Chicago. From August 2016 to June 2017, I also received fellowship funding from the Japanese Society for the Promotion of Science (JSPS) for study in Japan. During that fellowship, I was privileged to be affiliated with the Osaka University Department of Anthropology, where I was given opportunities to share my work with students and faculty and to participate in conferences and workshops.

With funding from the Center for East Asian Studies, my Japan studies colleagues and I had the privilege of hosting Japan research workshops at the University of Chicago in 2011 and 2012. Those workshops provided me with the first real opportunity to present early material from this book, and I am thankful for the response I received from the participants Andrea Arai, Kate Goldfarb, Junko Kitanaka, Phil Keffen, Koga Yukiko, Joseph Hankins, Gabriella Lukacs, David Novak, Lorraine Plourde, Satsuka Shiho, Watanabe Takehiro, and Takeyama Akiko.

I also want to thank Anne Allison. The third chapter of this book is developed from an article that I published in the *Journal of Cultural Anthropology* in 2013. Anne helped me greatly in refining the argument for that publication, and her influence has carried over into this book.

I am extremely grateful to Priya S. Nelson at the University of Chicago Press and her assistant Dylan Joseph Montanari for all the work they did to turn the manuscript into book form. Priya provided much needed encouragement early in the development of this book after reading two very early chapter drafts. I am thankful as well to Suzuki Wakana for chasing down and securing the images and permissions and to Kerry Higgins Wendt and Serene Yang for their marvelous copyediting. Abe Junko and her partner Satoshi were also invaluable in assisting me in the recovery of a number of difficult images.

Hyde Park was home from 2010 to 2016, and there are a great many friends and colleagues there who have taken an indirect role in helping me produce this book. I owe many thanks to Noa Viasman and Ofer Ravid for their cultural and academic support. Our neighbors Emil Sidky and Ingrid Reiser played an integral part helping look after our boys on weekends so that I could work and being forever ready to offer their expertise regarding physics and technical systems. Jason MacLean and Mayram Saleh were invaluable sources of friendship and conversation and were my experts in cybernetics, and Eugene Raikhel and Iris Bernblum provided friendship and support. Outside Hyde Park but close to home, thanks are due as well to Patrick and Maeve McWhinney.

In Japan, Nakamura Yutaka and his partner Ryoko provided much encouragement and were the source of many inspiring discussions on philosophy, social theory, and technology. I am also especially grateful to Chiba Hajime, who helped me find my way in Japan while I was on the JSPS Fellowship, and to the community of Maehama, especially Kikuchi Toshio, who provided a place for me to live while I worked on revisions.

My academic career began at McGill University many years ago with financial support from Mary and David Green. Mary and David showed great faith in me, which gave me the strength to pursue my studies at the time after many years outside academic life. They remain an inspiration.

I owe my deepest gratitude to family. Masaki and Yukiko Yanagita have been like family to me and my partner and our sons. They have been with me from the very first months of fieldwork for this book. I would not have made it through those years in Tokyo without the many long conversations into the night over elaborate dinners and drinks in their home. They are kindred spirits in every sense and have been intellectual mentors at every step of the way over the past years.

Hajime and Michiyo Mizukawa played a large role in the research for the last chapter of this book and provided support over the years. I am very grateful to Yuki Mizukawa for the constant and unconditional family support she has given over the years. Her home always provided a calm harbor to rest and to think.

My brothers and sisters Danny, Joel, Sharon, and Miriam and all their many sons and daughters have been sources of encouragement and inspiration. I am especially grateful to Danny for reading and commenting on chapter drafts and for his expert assistance in parsing many of the issues raised in this book.

I owe my greatest debt to my parents, Dr. James Fisch and Rochelle Fisch, whose love and support helped my family weather the years of my writing and whose weekly inquiry, "Nu, how's the book going?" I can finally answer.

Our sons Kai and Mio have supplied endless laughter and happiness to bring me through the writing process, and I continue to learn more about life from them every day. I wrote many sections of this book at odd hours of the night while bouncing gently up and down on a large exercise ball with Kai and Mio strapped to my chest when they were still infants. The rhythms of their breathing and heartbeats from those late hours are etched in these pages, informing and animating the text. Finally, to my partner in life and thought, Jun, who put her own academic career aside so that I could finish this book, I am forever grateful. She was involved with every part of this book. Moreover, she gave me the time, the courage, and the compassion I needed to write. I would not be where I am today without her.

Bibliography

Adam, Burgess, and Mitsutoshi Horii. "Constructing Sexual Risk: 'Chikan,' Collapsing Male Authority and the Emergence of Women-Only Train Carriages in Japan." *Health, Risk & Society* 14, no. 1 (2012): 41–55.

Adorno, Theodor W., Else Frenkel-Brunswik, Daniel J. Levinson, R. Nevitt Sanford, and the American Jewish Committee. *The Authoritarian Personality*. New York: Harper & Row, 1950.

Alexy, Allison. "Intimate Dependence and Its Risks in Neoliberal Japan." *Anthropological Quarterly* 84, no. 4 (2011): 895–917.

Allison, Anne. *Millennial Monsters: Japanese Toys and the Global Imagination*. Berkeley: University of California Press, 2006.

———. *Precarious Japan*. Durham, NC: Duke University Press, 2013.

Althusser, Louis. *On the Reproduction of Capitalism: Ideology and Ideological State Apparatuses*. 1971. Translated by G. M. Gosiigarian. Brooklyn: Verso, 2014.

Anand, Nikhil. "Pressure: The Politechnics of Water Supply in Mumbai." *Cultural Anthropology* 26, no. 4 (2011): 542–64.

Appel, Hannah. "Walls and White Elephants: Oil, Infrastructure, and the Materiality of Citizenship in Urban Equatorial Guinea." In *The Arts of Citizenship in African Cities: Infrastructures and Spaces of Belonging*, ed. Mamadou Diouf and Rosalind Fredericks. New York: Palgrave Macmillan, 2012.

Arai, Andrea. *The Strange Child: Education and the Psychology of Patriotism in Recessionary Japan*. Stanford, CA: Stanford University Press, 2016.

Ashby, W. Ross. "Principles of the self-organizing dynamic system." *The Journal of General Psychology* 37, no. 2 (1947): 125–28.

Augé, Marc. *In the Metro*. Translated by Tom Conley. Minneapolis: University of Minnesota Press, 2002.

———. *Non-Places: Introduction to an Anthropology of Supermodernity* 2nd English language ed. London: Verso, (1992) 2008.

Bataille, Georges. *The Accursed Share: An Essay on General Economy.* New York: Zone Books, 1988.

Beck, Ulrich. *Risk Society: Towards a New Modernity.* Translated by Mark Ritter. Theory, Culture & Society. Los Angeles: Sage Publications, 1992 (1986).

Benjamin, Walter. "The Work of Art in the Age of Mechanical Reproduction." In *Illuminations*, edited by Hannah Arendt, 217–52. New York: Schocken Books, 1968.

Berlant, Lauren Gail. *Cruel Optimism.* Durham, NC: Duke University Press, 2011.

Boellstorff, Tom. *Coming of Age in Second Life: An Anthropologist Explores the Virtually Human.* Princeton, NJ: Princeton University Press, 2008.

Bond, David. "Governing Disaster: The Political Life of the Environment During the BP Oil Spill." *Cultural Anthropology* 28, no. 4 (2013): 694–715.

Brockman, John R. *Twisted Rails, Sunken Ships: The Rhetoric of Nineteenth-Century Steamboat and Railroad Accident Investigation Reports, 1833–1879.* Amityville, NY: Baywood, 2004.

Brooks, Rodney A. "Intelligence without Reason." *The Artificial Life Route to Artificial Intelligence: Building Embodied, Situated Agents* (1995): 25–81.

———. "Intelligence without Representation." *Artificial Intelligence* 47 (1991): 139–59.

Buckley, Sandra. "Altered States: The Body Politics of 'Being-Woman.'" In *Postwar Japan as History*, edited by Andrew Gordon, 347–72. Berkeley: University of California Press, 1993.

Butler, Judith. *Bodies That Matter: On the Discursive Limits of "Sex."* London: Routledge, 1993.

Castells, Manuel. *The Rise of the Network Society.* Cambridge, MA: Blackwell Publishers, 1996.

Chu, Julie Y. "When Infrastructures Attack: The Workings of Disrepair in China." *American Ethnologist* 41, no. 2 (2014): 351–67.

Combes, Muriel. *Gilbert Simondon and the Philosophy of the Transindividual.* Translated by Thomas LaMarre. Cambridge, MA: MIT Press, 2013.

Cook, Haruko. "Meanings of Non-Referential Indexes: A Case Study of the Japanese Sentence-Final Particle Ne." *Test & Talk* 12, no. 4 (2009): 507–39.

Cooper, Melinda. *Life as Surplus: Biotechnology and Capitalism in the Neoliberal Era.* Seattle: University of Washington Press, 2008.

Dawkins, Richard. *The Selfish Gene.* Oxford: Oxford University Press, 1976.

Deleuze, Gilles. *Difference and Repetition.* New York: Columbia University Press, 1994.

———. "Postscript on the Societies of Control." *October* 59, Winter (1992): 3–7.

Deleuze, Gilles, and Félix Guattari. *A Thousand Plateaus: Capitalism and Schizophrenia.* Translated by Brian Massumi. Minneapolis: University of Minnesota Press, 1987.

Denkisha kenkyukai. *Kokutetsu Densha Hattatsushi.* Tokyo: Denkisha kenkyu-kai, 1959.

Douglas, Mary. *Purity and Danger: An Analysis of Concepts of Pollution and Taboo.* London: Routledge, 1966.

Downer, John. "'737-Cabriolet': The Limits of Knowledge and the Sociology of Inevitable Failure." *American Journal of Sociology* 117, no. 3 (November 7, 2011): 725–62.

———. "Trust and Technology: The Social Foundations of Aviation Regulation." *The British Journal of Sociology* (2010).

———. "When the Chick Hits the Fan: Representativeness and Reproducibility in Technological Tests." *Social Studies of Science* 37, no. 1 (February 1, 2007): 7–26.

East Japan Railway Company. "Company Data." www.jreast.co.jp/e/data/index.html?src=gnavi.

Edwards, Catharine, and Thomas Osborne. "Scenographies of Suicide: An Introduction." *Economy and Society* 32, no. 2 (2005): 173–77.

Egami, Setsuko. "Idō to seikatsu ni okeru aratana kachi no kōzō o mezashite—kachi · kaiteki · kukan no kōzō." *JR East Technical Review* 4 (2003).

Eguchi, Tetsuo. "Ressha Daiya Wa Ikimono." In *Tetsudō No Purofesshionaru,* edited by Hoshikawa Takeshi. Tokyo: Gakken, 2008.

Farberow, Norman L. *Suicide in Different Cultures.* Baltimore: University Park Press, 1975.

Fisch, Michael. *"Days of Love and Labor*: Remediating the Logic of Labor and Debt in Contemporary Japan." *positions: asia critique* 23, no. 3 (2015): 463–86.

———. "Tokyo's Commuter Train Suicides and the Society of Emergence." *Cultural Anthropology* 28, no. 2 (2013): 320–43.

———. "War by Metaphor in *Densha otoko.*" *Mechademia 4: War/Time* 4 (November 2009): 131–46.

Flaig, Paul, and Katherine Groo. *New Silent Cinema.* AFI Film Readers Series. New York: Routledge, 2016.

Fortun, Kim. *Advocacy after Bhopal: Environmentalism, Disaster, New Global Orders.* Chicago: University of Chicago Press, 2001.

———. "From Latour to Late Industrialism." *HAU: Journal of Ethnographic Theory* 4, no. 1 (2014): 309–29.

Foucault, Michel. *Discipline and Punish: The Birth of the Prison.* New York: Vintage Books, 1995.

———. *Security, Territory, Population: Lectures at the Collège de France, 1977–78.* Translated by Graham Burchell. Edited by Michel Senellart, François Ewald, and Alessandro Fontana Basingstoke. New York: Palgrave Macmillan, 2007.

4.25 Network and Nishinihon ryokaku tetsudō kabushikigaisha. *Fukuchiyama-sen ressha dassen jiko no kadai kentō kai hōkoku.* 2011.

Frasca, Gonzalo. "SIMULATION 101: Simulation versus Representation." Ludology.org. 2001. www.ludology.org/articles/sim1/simulation101.html.

Freedman, Alisa. "Commuting Gazes: Schoolgirls, Salarymen, and Electric Trains in Tokyo." *Journal of Transport History* 23, no. 1 (2002): 23–36.

———. *Tokyo in Transit: Japanese Culture on the Rails and Road*. Stanford, CA: Stanford University Press, 2011.

Fujii, James A. "Intimate Alienation: Japanese Urban Rail and the Commodification of Urban Subjects." *Differences: A Journal of Feminist Cultural Studies* 11, no. 2 (1999): 106–33.

Gad, Christopher, C. B. Jensen, and Brit Ross Winthereik. "Practical Ontology: Worlds in STS and Anthropology." *NatureCulture* 3 (2015): 67–86.

Galloway, Alexander, and Eugene Thacker. *The Exploit: A Theory of Networks*. Minneapolis: University of Minnesota Press, 2007.

George, Timothy S. *Minamata: Pollution and the Struggle for Democracy in Postwar Japan*. Harvard East Asian Monographs 194. Cambridge, MA: Harvard University Asia Center, distributed by Harvard University Press, 2001.

Gibson, William. *Pattern Recognition*. New York: G. P. Putnam's Sons, 2003.

Giddens, Anthony. *The Consequences of Modernity*. Stanford, CA: Stanford University Press, 1990.

Gitelman, Lisa. *Scripts, Grooves, and Writing Machines: Representing Technology in the Edison Era*. Stanford, CA: Stanford University Press, 1999.

Gleick, James. *Chaos: Making a New Science*. London: Vintage, 1998.

———. *The Information: A History, a Theory, a Flood*. New York: Pantheon, 2011.

Goddard, Michael. "Towards an Archaeology of Media Ecologies: 'Media Ecology,' Political Subjectivation and Free Radios." *The Fibreculture Journal: Digital Media + Networks + Transdisciplinary Critique*, no. 17 (2011): 6–17.

Graham, Stephen, and Simon Marvin. *Splintering Urbanism: Networked Infrastructures, Technological Mobilities and the Urban Condition*. New York: Psychology Press, 2001.

Guattari, Félix. *The Three Ecologies*. London: Athlone Press, 2000.

Halpern, Orit. *Beautiful Data: A History of Vision and Reason since 1945*. Durham, NC: Duke University Press, 2014.

Haraway, Donna. *When Species Meet*. Minneapolis: University of Minnesota Press, 2008.

Hardt, Michael, and Antonio Negri. *Multitude: War and Democracy in the Age of Empire*. London: Penguin, 2004.

Harootunian, Harry D. *Overcome by Modernity: History, Culture, and Community in Interwar Japan*. Princeton, NJ: Princeton University Press, 2000.

Harvey, David. *A Brief History of Neoliberalism*. Oxford: Oxford University Press, 2005.

Harvey, Penny, and Hannah Knox. *Roads: An Anthropology of Infrastructure and Expertise*. Expertise: Cultures and Technologies of Knowledge. Ithaca, NY: Cornell University Press, 2015.

Hashimoto, Takehiko, and Shigehisa Kuriyama, eds. *Kindai nihon ni okeru tetsudō to jikan ishiki* [Railway systems and time consciousness in modern Japan]. Chikoku no tanjō: kindai nihon ni okeru jikan ishiki no keisei. Tokyo: Sangensha, 2001.

Hatamura, Yōtarō. *Sōteigai o sōteiseyo!: Shippaigaku kara no teigen.* Tokyo: NHK Shuppan, 2011.

Hayles, Katherine. *How We Became Posthuman: Virtual Bodies in Cybernetics, Literature, and Informatics.* Chicago: University of Chicago Press, 1999.

Headley, Lee A. *Suicide in Asia and the Near East.* Berkeley: University of California Press, 1983.

Hein, Laura E. "Growth Versus Success: Japan's Economic Policy in Historical Perspective." In *Postwar Japan as History*, edited by Andrew Gordon, 99–122. Berkeley: University of California Press, 1993.

Horiguchi, Sachiko. "Hikikomori: How Private Isolation Caught the Public Eye." In *A Sociology of Japanese Youth: From Returnees to NEETs*, edited by Roger Goodman, Yuki Imoto, and Tuukka H. I. Toivonen. New York: Routledge, 2012.

Horii, Mitsutoshi. *Josei senyō sharyō no shakaigaku.* Tokyo: Shūmei Shuppankai, 2009.

Ichikawa, Kon, dir. *Man-in densha* [The full-up train]. 1957. Tokyo: Daiei Studios.

Ichikawa, Kon, and Yuki Mori. *Ichikawa Kon no eigatachi* [Films of Kon Ichikawa]. Tokyo: Waizu Shuppan, 1994.

Iga, Mamoru. *The Thorn in the Chrysanthemum: Suicide and Economic Success in Modern Japan.* Berkeley: University of California Press, 1986.

Iida, Yumiko. *Rethinking Identity in Modern Japan: Nationalism as Aesthetics.* London: Routledge, 2002.

Inoue, Yumehito. Interview with a Shinchōsha editor. Shinchōsha. Accessed March 16, 2017. www.shinchosha.co.jp/99/special/index.html.

———. *Kyūjūkyū nin no saishū densha* [99 Persons' Last Train]. Tokyo: Shinchōsha, 1996. Accessed October 25, 2008. www.shinchosha.co.jp/99/.

Isamu, Yoshitake. "Ressha shūchū seigyo sōchi (CTC) no kaihatsu." In *Tetsudō no purofesshionaru*, edited by Hoshikawa Takeshi, 132–36. Tokyo: Gakken, 2008.

Ishikawa, Tatsujiro, and Mitsuhide Imashiro. *The Privatisation of Japanese National Railways.* London: Athlone Press, 1998.

Ishitani, Seikan. *Kōgaku gairon* [Introduction to Engineering]. Tokyo: Coronasha, 1972.

Ito, Keiichi. "Development and Update of ATOS." *JR EAST Technical Review*, no 20, Summer (2011): 52–55.

Ito, Masami, and Yuasa Hideo. "Autonomous Decentralized System with Self-organizing Function and Its Application to Generation of Locomotive Patterns." Paper presented at the [First] Proceedings from the International

Symposium on Autonomous Decentralized Systems (ISADS), Kawasaki, Japan, March 30–April 1, 1993.

Ito, Mizuko, Daisuke Okabe, and Izumi Tsuji, eds. *Fandom Unbound: Otaku Culture in a Connected World*. New Haven, CT: Yale University Press, 2012.

Ito, Mizuko, Daisuke Okabe, and Misa Matsuda, eds. *Personal, Portable, Pedestrian: Mobile Phones in Japanese Life*. Cambridge, MA: MIT Press, 2005.

Ivy, Marilyn. "Formations of Mass Culture." In *Postwar Japan as History*, edited by Andrew Gordon, 239–58. Berkeley: University of California Press, 1993.

Jansen, K., and S. Vellema. "What Is Technography?" *NJAS—Wageningen Journal of Life Sciences* 57 (2011): 169–77.

Japan's Aircraft and Railway Accidents Investigation Commission. *Tetsudō jiko chōsa hōkoku sho: Nishinihon ryokaku tetsudō kabushikigaisha fukuchiyama-sen takarazuka eki amagasaki eki kan ressha dassenjiko*. Japan's Ministry of Land, Infrastructure, Transportation, and Tourism, 2007. http://www.mlit.go.jp/jtsb/railway/fukuchiyama/RA07-3-1-1.pdf.

Jasanoff, Sheila. "Beyond Calculation: A Democratic Response to Risk." In *Disaster and the Politics of Intervention*, edited by Andrew Lakoff, 14–41. New York: Columbia University Press, 2010.

———. *Learning from Disaster: Risk Management after Bhopal*. Philadelphia: University of Pennsylvania Press, 1994.

Jensen, Casper B. "Multinatural Infrastructure: Phenom Penh Sewage." In *Infrastructure and Social Complexity: A Routledge Companion*, edited by Penny Harvey, Casper B. Jensen, and Atsuro Morita, 115–27. London: Routledge, 2016.

———. "New Ontologies? Reflections on Some Recent 'Turns' in STS, Anthropology and Philosophy." *Social Anthropology/Anthropologie Sociale* 25, no. 4 (2017): 1–21.

Jensen, Casper Bruun, and Atsuro Morita. "Infrastructures as Ontological Experiments." *Engaging Science, Technology, and Society* 1 (2015): 81–87.

Johnson, Chalmers. *Miti and the Japanese Miracle: The Growth of Industrial Policy, 1925–1975*. Stanford, CA: Stanford University Press, 1982.

Johnston, John. *The Allure of Machinic Life: Cybernetics, Artificial Life, and the New AI*. Cambridge, MA: MIT Press, 2008.

JR nishi nihon anzen forōappu kaigi. *JR nishi nihon anzen forōappu kaigi hōkokusho* [The Report of the JR West Safety Follow-Up Meetings, referred to as the *Follow-Up Report*]. JR West, April 25, 2014. https://www.westjr.co.jp/safety/fukuchiyama/followup/pdf/followup_all.pdf.

Kainuma, Hiroshi. *Fukushima ron: genshi ryōku mura wa naze umaretaka*. Tokyo: Seidosha, 2011.

Kasai, Yoshiyuki. *Japanese National Railways: Its Break-Up and Privatization: How Japan's Passenger Rail Services Became the Envy of the World*. Kent, England: Global Oriental, 2003.

Kawashima, Ryōzō. *Naze fukuchiyama sen dassen jiko wa okotta no ka* [Why Did the Fukuchiyama Line Derailment Occur?]. Tokyo: Soshisha, 2005.

Kelly, William W. "Finding a Place in Metropolitan Japan." In *Postwar Japan as History*, edited by Andrew Gordon, 189–238. Berkeley: University of California Press, 1993.

Kirby, Lynne. *Parallel Tracks: The Railroad and Silent Cinema*. Durham, NC: Duke University Press, 1997.

Kitahara, Fumio, Kazuo Kera, and Keisuke Bekki. "Autonomous Decentralized Traffic Management System." Paper presented at the International Workshop on Autonomous Decentralized Systems, Chengdu, China, September 21–23, 2000.

Kitanaka, Junko. *Depression in Japan: Psychiatric Cures for a Society in Distress*. Princeton, NJ: Princeton University Press, 2011.

Kittler, Friedrich A. *Discourse Networks 1800/1900*. Stanford, CA: Stanford University Press, 1990.

Knight, Frank H. *Risk, Uncertainty and Profit*. New York: Houghton Mifflin, 1921.

Kuriyama, Shigehisa, and Takehiko Hashimoto, eds. *Chikoku no tanjō: kindai nihon ni okeru jikan ishiki no keisei* [*The Birth of Tardiness in Japan: The Formation of Time Consciousness in Modern Japan*]. Tokyo: Sangensha, 2001.

Kurosawa, Kiyoshi, dir. *Tōkyō sonata* [*Tokyo Sonata*]. 2008. Japan: Regent Releasing, Entertainment Farm; Fortissimo Films. DVD.

LaMarre, Thomas. "Afterword: Humans and Machines." Translated by Thomas LaMarre. In *Gilbert Simondon and the Philosophy of the Transindividual*, edited by Muriel Combes, 79–108. Cambridge, MA: MIT Press, 2013.

———. *The Anime Machine: A Media Theory of Animation*. Minneapolis: University of Minnesota Press, 2009.

———. "An Introduction to Otaku Movement." *EnterText* 4:1 (2004): 151–87.

———. "Living between Infrastructures: Commuter Networks, Broadcast TV, and Mobile Phones." *boundary 2* 42, no. 3 (2015): 157–70.

Langton, Christopher G., ed. *Artificial Life: An Overview*. Cambridge, MA: MIT Press, 1997.

Larkin, Brian. "The Politics and Poetics of Infrastructure." *Annual Review of Anthropology* 42 (Oct. 21, 2013): 327–43.

Latour, Bruno. *Reassembling the Social: An Introduction to Actor-Network-Theory*. Oxford: Oxford University Press, 2005.

———. *We Have Never Been Modern*. New York: Harvester Wheatsheaf, 1993.

LeCavalier, Jesse. *The Rule of Logistics: Walmart and the Architecture of Fulfillment*. Minneapolis: University of Minnesota Press, 2016.

Lefebvre, Henri. *The Production of Space*. Oxford: Blackwell, 1991.

Lévi-Strauss, Claude. *Totemism*. Boston: Beacon Press, 1962.

Lin, Zhongjie. *Kenzo Tange and the Metabolist Movement: Urban Utopias of Modern Japan*. New York: Routledge, 2010.

Lukacs, Gabriella. "Dreamwork: Cell Phone Novelists, Labor, and Politics in Contemporary Japan." *Cultural Anthropology* 28, no. 1 (2013): 44–64.

Mackenzie, Adrian. *Transductions: Bodies and Machines at Speed*. London: Continuum, 2006.

MacKenzie, Donald A., and Judy Wajcman. *The Social Shaping of Technology.* 2nd ed. Philadelphia: Open University Press, 1999.

Manovich, Lev. *The Language of New Media.* Cambridge, MA: MIT Press, 2002.

Martin, Reinhold. "The Organizational Complex: Cybernetics, Space, Discourse." *Assemblage* 37 (1998): 102–27.

Martinez, Antonio Garcia. *Chaos Monkeys: Obscene Fortune and Random Failure in Silicon Valley.* New York: HarperCollins, 2016.

Mazzarella, William. "Affect: What Is It Good For?" In *Enchantments of Modernity: Empire, Nation, Globalization,* edited by Saraubh Dube, 291–309. New York: Routledge, 2009.

———. "The Myth of the Multitude, or, Who's Afraid of the Crowd?" *Critical Inquiry* 36, no. 4 (Summer 2010): 697–727.

Meadows, Donella H., Dennis L. Meadows, Jorgen Randers, and William W. Behrens III. *The Limits to Growth.* New York: Universe, 1972.

Mindell, David A. *Between Human and Machine: Feedback, Control, and Computing before Cybernetics.* Baltimore: Johns Hopkins University Press, 2002.

Mito, Yuko. "Another 'Just in Time'—Japanese Significance." Hitachi-Rail.com. 2005. www.hitachi-rail.com/rail_now/column/mito/just_in_time2.html.

———. *Teikoku hassha: Nihon no tetsudō wa naze sekai de mottomo seikaku nanoka.* Tokyo: Shinchōsha, 2005.

Miyadai, Shinji, Yoshiki Fujii, and Akio Nakamori. *Shinseiki no riaru* [This century's new real]. Tokyo: Asuka Shinsha, 1997.

Miyamoto, Shoji, Kinji Mori, Hirokazu Ihara, Hiroshi Matsumaru, and Hiroyasu Ohshima. "Autonomous Decentralized Control and Its Application to the Rapid Transit System." *North Holland Computers in Industry* 5, no. 2 (1984): 115–24.

Mizukawa, Jun. "Reading 'on the Go': An Inquiry into the Temps and Temporalities of the Cellphone Novel." *Japanese Studies* 36, no. 1 (2016): 61–82.

Morita, Atsuro. "The Ethnographic Machine: Experimenting with Context and Comparison in Stratherian Ethnography." *Science, Technology & Human Values* 39, no. 2 (2014): 214–35.

———. "Multispecies Infrastructure: Infrastructural Inversion and Involutionary Entanglements in the Chao Phraya Delta, Thailand." *Ethnos* 0, no. 0,0 (2016): 1–20.

———. "Rethinking Technics and the Human: An Experimental Reading of Classic Texts on Technology." *NatureCulture* (2012): 40–58.

Morris-Suzuki, Tessa. *Beyond Computopia: Information, Automation, and Democracy in Japan.* London: Routledge, 1988.

———. *A History of Japanese Economic Thought.* Nissan Institute/Routledge Japanese Studies. London: Routledge/Nissan Institute for Japanese Studies, University of Oxford, 1989.

Murakami, Shōsuke, dir. *Densha otoko.* 2005. Tokyo: Toho Production Co., Ltd.

Nakamura, Hideo. *Ressha seigyo.* Tokyo: Kōgyō chōsa kai, 2010.

Nakano, Hitori. *Densha otoko.* Tokyo: Shinchōsha, 2004.

Nihon Kokuyū Tetsudō Sōsaishitsu Shūshika, ed. *Nihon kokuyū tetsudō hyaku-nenshi nenpyō.* 11 vols. Tokyo: Nihon kokuyū tetsudō, 1997.

Nishiyama, Takashi. "War, Peace, and Nonweapons Technology: The Japanese National Railways and Products of Defeat, 1880s–1950s." *Society for the History of Technology* 48, no. 2 (2007): 286–302.

Noda, Masaho, Harada Katsumasa, and Aoki Eichi. *Nihon no tetsudō: Seiritsu to tenkai.* Tokyo: Nihon Keizai Hyōronsha, 1986.

Nossal, G. J. V. *Antibodies and Immunity.* 2nd ed., rev. and expanded. New York: Basic Books, 1978.

Okazaki, Takeshi. "Besutoseraa: Shinsatsushitsu" [Bestseller: examination room]. *Chūōkōron* 120, no. 2 (February 2005): 264–65.

Osborne, Thomas. "'Fascinated dispossession': Suicide and the aesthetics of freedom." *Economy and Society,* 34:2 (2005): 280–94.

Parikka, Jussi. *Insect Media: An Archaeology of Animals and Technology.* Minneapolis: University of Minnesota Press, 2010.

Perrow, Charles. *The Next Catastrophe: Reducing Our Vulnerabilities to Natural, Industrial, and Terrorist Disasters.* Princeton, NJ: Princeton University Press, 2011.

———. *Normal Accidents: Living with High-Risk Technologies.* Princeton, NJ: Princeton University Press, 1999.

Pfaffenberger, Bryan. "Social Anthropology of Technology." *Annual Review of Anthropology* 21 (1992): 491–516.

Pickering, Andrew. *The Cybernetic Brain: Sketches of Another Future.* Chicago: University of Chicago Press, 2011.

———. *The Mangle of Practice: Time, Agency, and Science.* Chicago: University of Chicago Press, 1995.

Pinguet, Maurice. *Voluntary Death in Japan.* Cambridge, UK: Polity Press, 1993.

Reuleaux, Franz. *The Kinematics of Machinery: Outlines of a Theory of Machines.* Translated by Alex B. W. Kennedy. London: Macmillan, 1876.

Robertson, Jennifer Ellen. *Takarazuka Sexual Politics and Popular Culture in Modern Japan.* Berkeley: University of California Press, 1998.

Ryman, Geoff. *253: The Print Remix.* 1st St. Martin's Griffin ed. New York: St. Martin's Griffin, 1998.

Saito, Tamaki, and Jeffrey Angles. *Hikikomori: Adolescence without End.* Minneapolis: University of Minnesota Press, 2013.

Sand, Jordan. *Tokyo Vernacular: Common Spaces, Local Histories, Found Objects.* Berkeley: University of California Press, 2013.

Sarti, Alessandro, Federico Montanari, and Francesco Galofaro. *Morphogenesis and Individuation.* Lecture Notes in Morphogenesis. Cham, Switzerland: Springer, 2014.

Sasaki-Uemura, Wesley Makoto. "Competing Publics: Citizens' Groups, Mass Media, and the State in the 1960s." *positions: east asia cultures critique* 10, no. 1 (2002): 79–110.

Satō, Mitsuru. *Dare mo kataritagaranai: Testudō no uramenshi* [What nobody will talk about: behind scenes history of the railroad]. Tokyo: Saizusha, 2015.

———. *Tetsudō gyōkai no ura banashi.* Tokyo: Saizusha, 2014.

Sato, Yūichi. *Jinshin jiko dēta bukku 2002–2009.* Tokyo: Tsugeshobo, 2011.

Sawa, Kazuya. *Nihon no tetsudō koto hajime.* Tokyo: Tsukiji-shokan, 1996.

Schivelbusch, Wolfgang. *The Railway Journey: The Industrialization of Time and Space in the 19th Century.* Berkeley: University of California Press, 1986.

Schüll, Natasha Dow. *Addiction by Design: Machine Gambling in Las Vegas.* Princeton, NJ: Princeton University Press, 2012.

Sheller, Mimi. "Mobile Publics: Beyond the Network Perspective." *Environment and Planning D: Society and Space* 22 (2004): 39–52.

Shershow, Scott Cutler. *The Work and the Gift.* Chicago: University of Chicago Press, 2005.

Shneidman, Edwin S. *Comprehending Suicide: Landmarks in 20th-Century Suicidology.* Washington, DC: American Psychological Association, 2001.

Shosen Denshashi Kōyō. Tokyo: Tokyo tetsudō kyoku densha gakari tetsudo shi shiryo hozonkai hen, 1976.

Shosen Denshashi Kōyō. Tokyo: Tokyo tetsudokyoku densha gakari tetsudoshi shiryo hozonkai hen, 1927.

Simmel, Georg. "The Metropolis and Mental Life." In *The Sociology of Georg Simmel*, edited by Kurt H. Wolff, 409–26. Glencoe, IL: Free Press, 1950.

Simondon, Gilbert. "The Genesis of the Individual." Translated by Mark Cohen and Sanford Kwinter. In *Incorporations*, edited by Jonathan Crary and Sanford Kwinter, 297–319. New York: Zone, 1992.

———. "On the Mode of Existence of Technical Objects." *Deleuze Studies* 5, no. 3 (2011 [1958]): 407–24.

———. *On the Mode of Existence of Technical Objects.* Translated by Cecile Malaspina and John Rogove. Minneapolis: Univocal Press, 2017.

———. "Technical Mentality." Translated by Arne De Boever. In *Gilbert Simondon: Being and Technology*, edited by Arne De Boever, Alex Murray, Jon Roffe, and Ashley Woodward, 1–15. Edinburgh: Edinburgh University Press, 2012.

Sono, Sion, dir. *Jisatsu sākuru.* 2002. Tokyo: Kadokawa-Daiei Pictures.

Steger, Brigitte. "Negotiating Gendered Space on Japanese Commuter Trains." *Electronic Journal of Contemporary Japanese Studies* 13, no. 3 (2013). http://www.japanesestudies.org.uk/ejcjs/vol13/iss3/steger.html.

Suzuki, Atsufumi. *Densha Otoko wa dare nanoka: "Netaka" suru komyunikeishon* [Who is Densha Otoko: Neta-Ization communication]. Tokyo: Chūōkōron Shinsha, 2005.

Suzuki, Hiromi, and Tesuo Yamaguchi. *JR nishi nihon no tazai: Hattori untenshi jisatsu jiken to amagasaki dassen jiko* [JR West's great crime: the suicide of the driver Hattori and the Amagasaki derailment]. Tokyo: Gogatsu Shobo, 2006.

Takahashi, Yoshitomo. *Chūkōnen jisatsu: Sono jittai to yobō no tameni* [Middle and old age suicide: how to prevent that reality]. Tokyo: Mimatsu, 2003.

Takeuchi, Kaoru. *"Seimei ni manabu: Jiritsu bunsan toiu shisō – shikō o kaeru, shakai o kaeru, paradaimu o kaeru."* In *Hitachi hyōron,* 4–9. 2009.

Tanaka, Daisuke. "Shanai kūkan to shintai gihō. Senzenki densha kōtsū ni okeru 'waizatsu' to 'kōkyōsei'" ["Body Techniques in a Train: Publicity and Deviation in Urban Traffic During the Prewar Period"]. *Shakaigaku hyōron* 58, no. 1 (2007–2008): 40–56.

Tanaka, Stefan. *New Times in Modern Japan.* Princeton, NJ: Princeton University Press, 2004.

Thrift, Nigel. "Movement-space: The changing domain of thinking resulting from the development of new kinds of spatial awareness." *Economy and Society* 33, no. 4 (Nov 1, 2004): 582–604.

Tomii, Norio. *Resshya daiya no himitsu: Teiji unkō no shikumi.* Tokyo: Seizandō, 2005.

Tsutsui, William M. *Manufacturing Ideology: Scientific Management in Twentieth-Century Japan.* Princeton, NJ: Princeton University Press, 1998.

Turkle, Sherry. *Alone Together: Why We Expect More from Technology and Less from Each Other.* New York: Basic Books, 2011.

———. *Simulation and Its Discontents.* Cambridge, MA: MIT Press, 2009.

Vannini, Philip, Jaigris Hodson, and April Vannini. "Toward a Technography of Everyday Life: The Methodological Legacy of James W. Carey's Ecology of Technoculture as Communication." *Cultural Studies—Critical Methodologies* 9, no. 3 (2009): 462–76.

Wark, McKenzie. *Gamer Theory.* Cambridge, MA: Harvard University Press, 2007.

Watanabe, Kōhei, and Hiroshi Tamura. *Ryojō 100–nen: Nihon no tetsudō.* Tokyo: Mainichi Shinbunsha, 1968.

Watanabe, Susumu. "Restructuring of Japanese National Railways: Implications for Labour." *International Labour Review* 133, no. 1 (1994): 89–111.

Weathers, Charles. "Reconstruction of Labor-Management Relations in Japan's National Railways." *Asian Survey* 34, no. 7 (July 1994): 621–33.

West, Mark D. *Law in Everyday Japan: Sex, Sumo, Suicide, and Statutes.* Chicago: University of Chicago, 2005.

Wiener, Norbert. *The Human Use of Human Beings: Cybernetics and Society.* London: Free Association, 1954.

Wigley, Mark. "Network Fever." In *New Media, Old Media: A History and Theory Reader,* edited by Wendy Hui Kyong Chun and Thomas Keenan, 375–98. New York: Routledge, 2005.

Yamagishi, Toshio. *Shinrai no kōzō: Kokoro to shakai no shinka gēmu.* Tokyo: Tōkyō Daigaku Shuppankai, 1998.

Yamamoto, Hirofumi. *Technological Innovation and the Development of Transportation in Japan.* Tokyo: Unipub, 1993.

Yamamoto, Masahito. "Sekai ni hirogaru jiritsu bunsan." *Landfall* 48 (2003): 1–5.

Yaneva, Albena. "Scaling Up and Down: Extraction Trials in Architectural Design." *Social Studies of Science* 35, no. 6 (2005): 867–94.

Yoda, Tomiko. "A Roadmap to Millennial Japan." *South Atlantic Quarterly* 99, no. 4 (2000): 629–68.

Ziman, J. M. *Technological Innovation as an Evolutionary Process*. Cambridge, UK: Cambridge University Press, 2000.

Žižek, Slavoj. *For They Know Not What They Do: Enjoyment as a Political Factor*. Radical Thinkers. London: Verso, 2008.

Index

Page numbers in italics refer to figures and photos.